United States Women in Aviation 1919–1929

Kathleen Brooks-Pazmany

SMITHSONIAN INSTITUTION PRESS
Washington and London

Library of Congress Cataloging in Publication Data
Brooks-Pazmany, Kathleen L.
United States women in aviation, 1919–1929.
Bibliography: p.
1. Women air pilots—United States. I. Title. II.
Series. TL539.B76 1982 629.13'092'2 82-10812
ISBN: 0-87474-378-8

Manufactured in the United States of America
97 96 95 94 93 5 4 3 2

♾ The paper used in this publication meets the minimum requirements of the American National Standard for Permanence of Paper for Printed Library Materials Z39.48-1984

CONTENTS

United States Women in Aviation 1919–1929

Kathleen L. Brooks-Pazmany

Introduction

The challenges that confronted women pilots in the 1920s are faced by all women entering a field that is generally considered a masculine domain. The question of whether aviation was or is a "proper" sphere for women is still unresolved in many minds. Despite the odds against them, women of that era made great contributions to the development of aviation. Women of today in any field can look to their experiences for inspiration.

We are all familiar with the names of Amelia Earhart, Anne Morrow Lindbergh, and Jacqueline Cochran, but for most of us that is where our knowledge of women in aviation ends. Yet from the beginning women have been an integral part of aviation's progress. Katherine Wright devoted a great deal of time and effort to her brothers' research although she did not fly herself. In fact, when her brothers opened a flying school in Montgomery, Alabama, in 1910, Orville rejected all the female applicants on the grounds that they were notoriety seekers. Katherine believed in the future of aviation, and her brother's attitude did not stop her from contributing to its progress.

The attitude that flying was socially inappropriate and even physically impossible for women was common. To overcome the prejudice and the overprotectiveness, women often took to the skies first as passengers before assuming a more visible role as pilot, thereby demonstrating the relative ease and safety of air travel. Lillian Gatlin, the first woman to make a transcontinental flight, was a passenger. In the early attempts to cross the oceans women flew as passengers rather than pilots. But even as passengers they were taking risks, and some lost their lives.

There were only a handful of qualified woman pilots at the end of World War I. During the twenties an ever-increasing number of women barnstormed, raced, carried passengers, and set records. By the end of the decade they had helped to bring the airplane before the public as a practicable mode of transportation and a wonderful new form of recreation. Their contributions argue for the recognition of a place for both men and women in this and all fields of human endeavor.

Much of the material used in this publication came from the National Air and Space Museum Library. This library contains an extensive collection of publications, photographs, and other documentation on the history of aviation. Yet by far some of the most significant information came from the participants themselves; Bobbi Trout, Viola Gentry, Elinor Smith, Mary Haizlip, Nellie Zabel Willhite, Martin Jensen, Dexter Martin, and Lillian Boyer. They added invaluable insight to the events of the day, provided information and photographs, and guided me to other sources. Glenn Buffington, Babe Weyant Ruth, and Rose LeDuke have also

Kathleen L. Brooks-Pazmany, Department of Aeronautics, National Air and Space Museum, Smithsonian Institution, Washington, D.C. 20560.

shared their personal collections and knowledge with me. Without their help this study would not have been possible. I would also like to offer special thanks to Louise Thaden for her quick and generous response at the outset of my research; I regret that she never saw the finished product. Note should also be made of those on the Museum staff who contributed their time and talent to this study, especially to Dale Hrabak and the photo lab staff; Susan Brown, formerly of the Aeronautics Department; Dorothy Cochrane; and Susan Owen.

Although the emphasis here is on the most visible figure in aviation, the pilot, the many women who filled support roles should not be forgotten; there were more of them than ever found their names in print. Throughout the period from 1919 to 1929 there was a rapid increase in the number of women interested in aviation as well as a greater opportunity for them to become actively involved. Over the years it became socially more acceptable, at least for a single woman, to pursue a career in aviation. With the advances in aircraft design and maneuverability, the physical barriers to women flying were greatly reduced; strength and size were no longer so important. Now was the time to move forward.

Women Take a Place In Aviation

> In the history of the conquest of the air the page of women's achievement is a great one. The woman of today is something more than a wife to her husband. She is his friend, his companion, his playmate, and more often than not his business partner. The true development of the air sense of mankind depends just as much on the women as on the men. We cannot advance with a united front into the position of leadership in this new world that is ours by right, unless the women will support and help men as they have always done in the past. Since the beginning of human flight women have stood by men not alone in actual flying, but in the development of aeronautics generally.
>
> (Heath, 1929:32)[1]

When civilian flying in the United States was curtailed during World War I, women pilots found various ways to stay active in aviation. Marjorie Stinson took a position as an aeronautical draftsman for the United States Navy Department in 1918. Anita "Neta" Snook also kept her hand in aviation during the war. She began her flight training in 1917 in Davenport, Iowa. When the Davenport Aviation School closed after an accident in which a potential investor was killed and the instructor critically injured, Neta went to Newport News, Virginia, and continued her training at the Curtiss school there. Still she had not soloed when the wartime ban was placed on civilian flying. If she could not fly for the war effort, then she could put her experience in building and maintaining airplanes to good use. Neta took a position as expeditor for the British Air Ministry, monitoring and inspecting aircraft engines under production at the Willys Morrow Factory in Elmira, New York.

It was as a result of a Liberty Loan drive that Laura Bromwell took her first airplane ride. When she heard that the Virginian with the best sales record for Liberty Bonds would be given a ride in an airplane, she set out to win that flight by selling $21,000 worth of bonds.

When the ban on civilian flying was lifted in 1919, Laura wasted little time before she began taking lessons, and on 22 October of that year she became the first woman after the war to receive her pilot's license from the Aero Club of America. By coincidence, the following day an article appeared in the *New York Times* announcing "WOMEN AVIATORS WANTED—Police Reserves Seeking Recruits for the School now in Operation" (author unknown, 1919c:17).

The New York Aerial Police Reserves were seeking women recruits between the ages of 18 and 25 for the Women's Aviation Corps, which would be attached to the Women's Police Reserves. Candidates for the Corps had to pass both physical and mental examinations before being accepted. They also had to purchase their own uniforms and books for the course of instruction. The Corps was a volunteer organization, with several of its pilots providing their own airplanes. The group was under the financial sponsorship of Rodman Wanamaker, Special Deputy Policy Commissioner,[2] who hoped that there would eventually be a permanent corps of men and women to perform such duties as regulating

[1] Though Lady Mary Heath was not a U.S. citizen she was an avid proponent of the development of aviation in this country. She spent several years on the lecture circuit, participating in aviation events like the 1929 National Air Races, and lobbying for aviation legislation in Washington, D.C.

[2] Rodman Wanamaker held this position by virtue of the fact that he was a millionaire who had joined the New York Police Department without taking a salary. He was one of several men to hold what could be construed as honorary positions, and he functioned as an advisor to the police force.

FIGURE 1.—Marjorie Stinson at her drafting table at the U.S. Navy Department of Aeronautic Design. (Courtesy of Marjorie Stinson Collection, Library of Congress)

and controlling traffic, transporting prisoners, and pursuing criminals. With this in mind he became the principal advocate for the New York Aerial Police in the early months of 1918. During World War I, Wanamaker had established a flying school to train military pilots. He planned to use his facility to train pilots for the aerial police squadron. For the initial establishment of the squadron however, he was counting on returning Army pilots to "enlist." Even though it was a reserve force with no regular pay, there was no shortage of applicants.

The squadron's organization closely followed U.S. Army requirements, the idea being that if another war broke out a reserve of trained personnel would be available. Some aircraft were provided by reservists, and the Navy lent the squadron four hydroplanes. Wanamaker invested his own money to maintain and promote the aerial squadron, but he was never able to obtain financial support from city officials. In 1922, because of a lack of funds, members of the aerial reserve were mustered into the Navy Reserves. They still maintained their status as Police Reserves and could be called on in emergencies. The Women's Aviation Corps had already disappeared by this time, but in the interim, they had performed at local aerial exhibitions to promote "airmindedness" in the community by showing spectators how safe and how utilitarian airplanes were.

Though the Corps was a volunteer body and its members were not paid for their efforts, Laura Bromwell enlisted and was thus able to fly at no expense to herself. She received a commission as lieutenant and soon drew wide attention for her aerial performances. On 20 August 1920, during

FIGURE 2.—Laura Bromwell smiles triumphantly after setting a new women's record for looping the loop. (SI photo 80-452)

rededication ceremonies of Curtiss Aerodrome (formerly Hazelhurst Field), Mineola, New York, the astounded crowd saw her loop-the-loop 87 times, breaking Adrienne Bolland's record of 25 consecutive loops made on 12 April 1920 in France. Laura had counted over 100 loops, but the clouds obscured her plane from the view of the observers during part of the flight.

In addition to her volunteer flying, Laura became the first woman to make a commercial flight over New York City. On 17 February 1920, she flew a Curtiss Oriole with "FLY HIGH! with LOCKLEAR in *The Great Air Robbery*" painted on its side and dropped leaflets over the theater district announcing the New York opening of this Universal motion picture.

She continued to excel in exhibition flying. On 15 May 1921, she set yet another women's loop-the-loop record. Before a crowd of 10,000 gathered for the opening of the Field Club of the Aero Club of America at Curtiss Field, Garden City, Long Island, she performed 199 consecutive loops in a Curtiss J-1, stopping only when she ran low on fuel.

Laura had great plans; she wanted to establish a new altitude record and to be the first woman to fly across the continent. But on 5 June 1921, just three weeks after her record-breaking flight, she died in a crash of a Curtiss Canuck. She had been cautioned not to stunt in the plane as she had not used it previously. While looping, Laura lost control of the plane and crashed. An investigation of the wreckage revealed no mechanical failure of the airplane, but a cushion was found on the roof of the Curtiss factory. This led investigators to the conclusion that she had not fastened her seat belt properly and when the cushion fell out, she could not reach the foot controls while upside down. Her death caused some outcry, but a New York Times editorial gave a different view of the accident (author unknown, 1921:16):

> So many men now have lost their lives in airplane accidents that individual addition [sic] to the long list of their names have ceased to cause any really deep emotions except in the minds of their relatives and friends. When a woman is the victim however the feeling of pity and horror is as strong as was that produced by the first of these disasters to men and though there is at present no expectation that aviation should be abandoned by men because of the recognized dangers, the death of Miss Bromwell is almost sure to raise in many minds at least the question if it would not be well to exclude women from a field of activity in which there [sic] presence certainly is unnecessary from any point of view.
>
> That question is worthy of consideration, but it should not be answered hastily, and especially its answer should not be based on this particular accident, lamentable as it was. There is little if any reason for assuming that Miss Bromwell as an aviator was less competent than a man of the same training, experience and ambition or taste. She had made many flights safely, and though there seems to be some evidence of something like carelessness in her preparations for her last flight—some failure to take all possible precautions, that is—there was nothing characteristic of her sex in that—nothing unlike what many male aviators have done sometimes with like results.
>
> All fliers are so wonted to the taking of risks and so frequently do they go aloft when conditions are not perfect and the principle of "safety first" would keep them on the ground that Miss Bromwell's failure to have her holding strap in perfect order should not be taken as counting against women as a class to compete with men in this new profession.

After the war, when flying was again open to civilians, there was great competition for aviation-related positions. During the war, airplane production had blossomed, but with its end hundreds of military aircraft were declared surplus. Prices of airplanes dropped drastically, making them affordable for many American pilots. But with the glut of aircraft on the market, design and production of new aircraft slumped and fewer jobs were available in the factories. The job market was flooded with pilots and mechanics just released from the armed

FIGURE 3.—Helen Lach waves to onlookers from the wing of the plane before her performance. Her parachute is at her feet. (SI photo 79-14589)

FIGURES 4, 5.—Lillian Boyer barnstormed through the twenties, performing a wide variety of stunts. (Courtesy of Lillian Boyer Werner)

forces and looking for work in aviation. It was not unusual to pick up a newspaper or magazine and find under "Situation Wanted" such items as the following (author unknown, 1919a:30):

AVIATOR discharged from active army service desires position. Experienced in auto business and accounting work. Prefer connection where services of a pilot are required. Address H, Box 5, Ace Publishing Co.

AVIATION MECHANIC, 14 months service, desires position with commercial aviation company. Best of references. Address N, Box 7, Ace Publishing Company.

With so few aviation positions available, many pilots went into business for themselves as barnstormers. Popular magazines, including *The Ace*, encouraged such enterprise in articles like the following, entitled "Stunt Pilots—Incorporated" (author unknown, 1919b:16):

Are you looking for a new profession, a new investment? Here is an opportunity—we suggest. Any number of army pilots have received their discharges from the aviation service. Most of them burn with desire to return to regions where they may continue their aerial pasttimes [sic].

All of the public is interested in aviation. Many individuals want to buy their own airplanes. County and state fairs, carnivals, circuses, and gala occasions of all sorts eagerly utilize stunt aviators. Moving picture directors are finding fresh themes—or may find fresh themes—in the field of aviation.

Some enterprising pilots, discharged army aviators, or perhaps some enterprising capitalist may believe us, and organize or cooperate, when we state that there is real opportunity for an incorporated company, which wishes to advertise nationally, to come into the market with stunt pilots, exploring pilots, photographic pilots, or experimenting pilots, as the goods which they have to market.

A telegram would come in—as a result of publicity. Fort Dodge, Iowa, will pull an agricultural show. Will Stunt Pilots, Inc., forward at once one knock-you-dead aviator, with plane, for entertainment purposes.

Some researchful engineer wants a photograph of the top of a brush-covered mountain. Will Stunt Pilots, Inc., forward one map making aviator?

It is too bad to waste the knowledge of aviation which various pilots, who cannot afford to purchase planes of their own, possess. It is too bad to leave these young men discontentedly working at some office work, or some dull profession, when their hearts are all in aviation, and aviation is growing—growing—growing.

Aeronautical progress needs their services; they need aerial opportunities. Science needs the impetus which opportunity granted them, they will give to aviation.

For women, too, barnstorming was often the only means of access to a career in aviation. They generally began by performing as wing walkers and parachutists. In 1927, 21-year-old Helen Lach of Wallington, New Jersey, was working her way through secretarial school by waiting on tables when she decided that parachute jumping at aerial exhibitions would be a more exciting way of earning

FIGURE 6.—Lillian Boyer performs a car-to-plane transfer in St. Paul, Minnesota, in 1922. (Courtesy of Lillian Boyer Werner)

money. When interviewed by the press about her jumps, Helen responded, "I get a wonderful thrill as I sail down to earth from the clouds, or I wouldn't take such a chance" (author unknown, 1927d:2).

Lillian Boyer was also looking for a career that would take her away from offices and restaurants and into the open air. In 1921, she was working as a waitress in a Chicago restaurant when two customers offered to take her for an airplane ride. She accepted and took her first flight on 3 April. On her second flight, on 7 April, she climbed out on the wing. Her career as an aerial exhibitionist had begun. On 10 October 1921 she made her first plane-to-plane change for the *International News*. In December of that same year she signed a contract with Lt. Billy Brock, former World War I pilot and barnstormer, and went to Chattanooga, Tennessee for five months of intensive training before taking the show on the road. During her career, which lasted until 1929 when federal regulations on low flying forced her and many other barnstormers into retirement, Lillian performed in 352 shows in 41 states and Canada. She made 143 auto-to-plane changes, 37 parachute jumps (13 of which landed the nonswimmer in Lake Erie), besides the usual repertoire of wingwalking stunts.

Many of these performers had hopes of becoming pilots as well. Such was the case with Phoebe Fairgrave. When she finished high school in 1920 she was eager to learn to fly, but no one would take her seriously. Phoebe was not easily discouraged (Orr, 1935:14):

> It was not long until I began to realize that one way in which I certainly could get up in the air would be to buy an airplane. I had a hunch that the prospect of a sale would make the boys waver in their determination not to have anything to do with satisfying my ambition to go aloft.

She took a small legacy and "shot the works." Her strategy worked. But flying lessons and upkeep on her plane were expensive, and the only way she could see of raising the money was by learning how to wingwalk and to make parachute jumps.

Phoebe made her first jump on 17 April 1921, and was left dangling unhurt in the trees. She soon joined the Glenn Messer Flying Circus and barnstormed throughout the Midwest. On 10 July 1921, Phoebe established a new world's record for women when she jumped from 15,200 feet. This jump made her one of the most popular attractions of the airshow. She enhanced her reputation by perfecting double parachute jumps. After the first chute opened she would cut it loose, free-falling for some moments before releasing the second chute. By the time she was twenty she became the first woman to form her own flying troupe, The Phoebe Fairgrave Flying Circus. Vernon Omlie, the man who taught her to fly, was her chief pilot; they were married in 1922. For Phoebe and Vernon, barnstorming was a means to an end. They continued to perform until they had saved enough money to start their own flying school and aviation business, Mid-South Airways in Memphis, Tennessee.

Barnstormers traveled around the country, thrilling spectators with death-defying aerial feats and selling airplane rides. It was very much a hand-to-mouth existence for most of them. And if it took dedication to fly and resilience to weather the financial hardships of such a life, it certainly took courage and determination to challenge the skies with a hearing impairment. Nellie Zabel Willhite had that courage and determination. Despite the fact that measles had left her virtually deaf, she resolved to learn to fly. On 13 January 1928 Nellie soloed, and thus became South Dakota's first licensed woman pilot. Her father bought her an Alexander Eaglerock which she named "Pard." Nellie became a familiar sight at airshows and fairs throughout the Midwest as the only woman pilot in the area. She flew in races and participated in flour bombings, in which pilots dropped bags of flour on ground targets. One of the most difficult events she took part in was balloon racing. In this event pilots would try to fly

FIGURE 7.—Despite the fact that she was deaf, Nellie Willhite obtained her license and performed at airshows for a number of years. (Courtesy of N.Z. Willhite)

FIGURE 8.—Margie Hobbs, a popular barnstormer who performed under the name "Ethel Dare," poses on the wing of her plane. (Courtesy of Rose LeDuke)

into balloons floating in the air. It required a great deal of skill to make the sharp turns necessary to hit the balloons. Nellie did the usual promotional flying, taking paying customers up for a five to ten minute flight over the countryside. Although she drew a good deal of attention wherever she flew, Nellie was "lonesome" being the only woman pilot in South Dakota.

Being lonesome was not the only problem a woman had to face on the barnstorming circuit. Public opinion and political pressure had a great impact on the success or failure of a woman's aerial career. For Margie Hobbs the transition from circus trapeze artist to wingwalker and transfer artist had been very easy. Billed as Ethel Dare, "The Flying Witch," she was still in her teens when she made her first successful transfer in November 1919. She thrilled audiences in Illinois and Wisconsin. She was somewhat perplexed at the awe with which she was viewed. She once protested to reporters: "I am human and ordinary and all that. . . . Honest, I eat food and sleep with my eyes shut just like everybody."[3] Unfortunately, her aerial career came to an abrupt halt. On 6 September 1920 at an airshow in Detroit, Michigan, Myron L. "Fearless" Tinney had

[3] Newspaper clipping, Ethel Dare biographical file, Library, National Air and Space Museum.

FIGURE 9.—The skills Ethel Dare had acquired in a circus trapeze act were useful when she made the transition to barnstormer. (Courtesy of Dexter Martin)

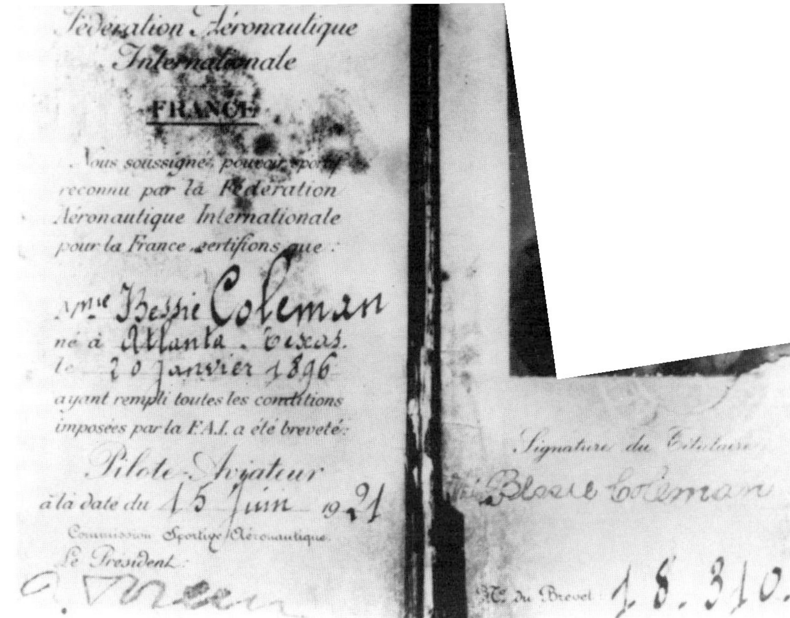

Fédération Aéronautique
Internationale
FRANCE

Nous soussignés pouvoir sportif reconnu par la Fédération Aéronautique Internationale pour la France certifions que:

Mlle Bessie Coleman
née à Atlanta, Texas.
le 20 Janvier 1896
ayant rempli toutes les conditions imposées par la F.A.I. a été breveté:
Pilote-Aviateur
à la date du 15 Juin 1921

Commission Sportive Aéronautique
Le Président:

Signature du Titulaire
Bessie Coleman

N° du Brevet 18.310.

FIGURES 10, 11.—Left, Bessie Coleman with Anthony Fokker, whom she met while she was taking flying lessons in France. Right, her FAI pilot's license, the first issued to a black woman. (Courtesy of Bessie Coleman Aviators)

been killed when he fell in an attempt to make a plane-to-plane transfer via a rope ladder suspended beneath one of the planes—the same stunt Ethel performed on a regular basis. The next day, at the same air show, Ethel made the transfer without any problem, but the following day Commissioner Couzens (later to become a Senator) put a ban on her flying. She was suddenly out of work. As she commented to the press, "Detroit's atmosphere is bad. . . . It's one of the first couzens of poverty."[4] She was unable to find work with another air circus, and eventually returned to vaudeville.

With the proliferation of small air circuses and teams of barnstormers, the public began to demand more thrills. They had become discriminating, paying only to see shows that offered them something out of the ordinary. Bessie Coleman had something unusual to offer the crowds. She was the first black woman from the United States to receive a pilot's license. If it was difficult for white women to break into aviation it was close to impossible for Bessie. Flying school after flying school turned her down, but she was determined to get a license. Finally, at the suggestion of Robert S. Abbott, editor and publisher of the *Chicago Weekly Defender*, Bessie went to France for flying lessons. In September 1921, she returned to the United States with her Federation Aeronautique Internationale (FAI) license in hand.

Bessie wanted to open a flying school so that other blacks would not experience the difficulties she had encountered. Realizing that it would take money, she too turned to barnstorming. The color of her skin, which had hindered her in her quest for flying lessons, now became her drawing card. She began in 1922 by giving shows in the Chicago area, and later moved her base of operations to Houston, Texas. She continued performing throughout the South, drawing large crowds of the curious, both black and white, wherever she went. Bessie was on the verge of retiring from barnstorming in April 1926; she had nearly enough money to open that flying school of which she had dreamed for so long.

On 30 April 1926, while preparing for May Day celebrations in Orlando, Florida, Bessie, accompanied by her mechanic and publicity agent, William Wills, took her plane up for a test flight. It was her last. When she put the plane into a nose dive it suddenly flipped over and Bessie, who had neither fastened her seat belt nor worn a parachute, was thrown from the plane and plunged to her death. Wills died with the impact of the crash. It would be many years before another black woman would gain a notable position in aviation.

Death was a risk every barnstormer faced and acknowledged, but tragedies did not always occur in the air. On 15 August 1927, Gladys Roy, who had made a name for herself by dancing the Charleston on the upper wing of a plane, died when she walked into the spinning propeller of a plane that was sitting on the ground.

Barnstorming did not always keep a pilot in the

[4] Ibid.

FIGURE 12.—Gladys Roy dancing the Charleston on the wing of her plane in flight. A radio receiver sits on the wing. (SI photo 79-13683)

FIGURE 13.—Ruth Law with Shigenobu Okuma, former Prime Minister of Japan, during her tour of the Pacific in 1919. (SI photo 79-14300, damaged)

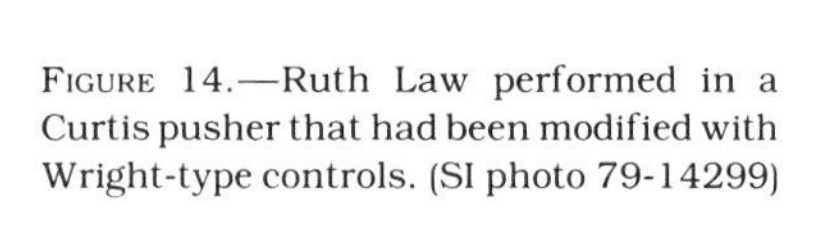

FIGURE 14.—Ruth Law performed in a Curtis pusher that had been modified with Wright-type controls. (SI photo 79-14299)

poorhouse. Some, such as Ruth Law, were quite successful. Before the outbreak of World War I Ruth had been making $9000 a week through commercial and demonstration flying. Immediately after the war she took her aerial exhibitions abroad, visiting Japan, China, and the Philippines before going to Europe. In Manila on 4 and 5 April 1919, Ruth Law became the first person to carry the airmail on an experimental flight sponsored by the Aero Club of the Philippines.

By the time she returned to the United States, barnstorming was quite the rage. Ruth and her husband/manager organized a three-plane flying troupe that they called Ruth Law's Flying Circus. Soon her circus was one of the most popular on the circuit. With Ruth in center position, the planes would fly in tight formation. She also piloted the plane for car-to-plane transfers. Her favorite stunt, however, was to race at low altitude with an automobile around a speedway or racetrack.

Ruth Law's aviation career came to an abrupt end. One morning in 1922 Ruth opened her newspaper to read an announcement that she would never fly again. Ruth never made such a declaration; her husband had done it without her knowledge. To appease the crowds, Ruth's stunts had become increasingly dangerous. In one of the last stunts she had perfected she climbed out of the cockpit, walked to the center of the biplane's wing, and stood upright while the pilot made three consecutive loops. Fearing for her life, her husband sent an announcement of her retirement to the newspapers. Since they did not need the money, she withdrew from flying in deference to her husband's wishes.

When Mabel Cody arrived on the southeastern airshow circuit in 1924, barnstomers throughout the area were given a run for their money. Her phenomenal skill and grace and the cool confidence with which she performed soon won her the admiration of the South.

It did not take Mabel long to realize that she would be much better off running her own show, and she soon made the Mabel Cody Flying Circus one of the most spectacular shows in the region. Mabel modeled her show after the circus, having three Sky Rings of performances. She performed some of the show's most spectacular feats, billing herself as the "Greatest Aerialist and Transfer Artist." She thrilled crowds with such stunts as wingwalking, dangling from the plane's spreader bar with one hand, and transferring from one plane to another.

One of her most spectacular stunts was her car-

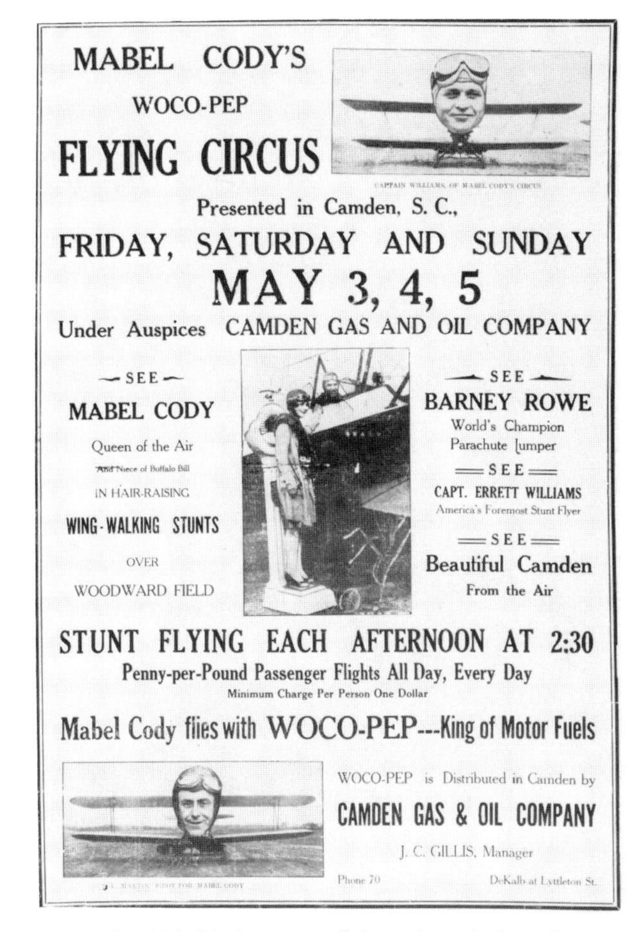

FIGURE 15.—Mabel Cody was one of the most popular barnstormers in the South. (SI photo 80-700)

to-plane transfer. As her car sped along a track a plane would approach, rope ladder extended. With one hand she would grab the last rung and be lifted into the air. As if endangered by the added weight, the plane would dip, giving the impression that it might crash. Then it would level off as Mabel clambered up the ladder.

Mabel continued to develop breathtaking feats. In April 1927 in St. Augustine, Florida, she used the same techniques to transfer from a Gold Cup speedboat racer, the *Miss Tampa*, to an East Coast Airways plane.

World War I had greatly stimulated the development of the airplane. Aircraft were now consistently capable of carrying heavy loads for sustained lengths of time, and could prove comparatively fast and safe transportation. At the end of the war there was a strong effort to maintain the momentum in the design and production of new aircraft, but this

FIGURE 16.—Mabel Cody's speedboat-to-plane transfer required both expert timing and good judgment. (SI photo 79-7192)

proved impossible. Americans were more interested in the "return to normalcy," so the infant aviation industry languished.

The most obvious and immediate way of keeping public interest in aviation alive had been through barnstorming. Many of the men and women who took up this risky profession viewed it as a stop gap measure—a way to keep flying and make some money until aviation got back on its feet. Unfortunately, the dangers inherent in barnstorming continued to reinforce the notion that flying was a costly and dangerous game for the reckless.

During the next few years women had a considerable influence in changing flying from a novelty to a lucrative commercial industry. Slats Rogers, a barnstormer of the time, reminisced about the feminine fascination with flying (Stilwell and Rogers, 1954:61):

> It was funny as hell watching the people go up for the first time. Here would come some middle-aged man and wife, and they would fight anybody who wanted to horn ahead of them, but they were scared white all the time.
>
> About the time it was their turn, the man would want to back out. But the woman would argue him into it. Almost every time it was that way—the man was the one that had to be talked into going up, the woman did the talking.

Women in aviation were still such a novelty in

FIGURE 17.—Lillian Gatlin beside the plane in which she made her coast-to-coast flight in 1922. (SI photo 79-11126)

the field that when a woman boarded an airplane it made news. The newspapers were just as likely to carry a story about Admiral Moffet's daughter Janet taking flying lessons as they were to report that Mrs. Frank Freeman had just taken her first airplane ride at the age of 85 and loved every moment of it. Both the industry and individuals were quick to capitalize on the press's eagerness to publicize the woman in aviation. Taking an airplane flight or buying a plane was sure to put any Hollywood starlet or entertainer's name in bold print. In September 1919, the *New York Herald* reported that Hope Eden, a perfomer in a vaudeville psychic act, had purchased a Curtiss JN-4D and hired a pilot. She explained that she hoped to avoid such problems as missing trains, losing baggage, and arriving late for an opening performance in another town. She probably drew more attention to herself this way than through advertising her act, and aside from the cost of the plane, it was free publicity.

Women found that they could also attract attention to personal causes through air travel. Lillian Gatlin, President of the National Association of Gold Star Mothers, became the first woman to cross the continent by air. She made the flight as a special delivery package in a mail plane because there was no commercial passenger service. She hoped that the flight would create interest in having the second Sunday in March set aside as a Memorial Day to Fliers.[5]

She flew under the auspices of Paul Henderson, Assistant Postmaster General, in a U.S. Post Office de Havilland mail plane, which followed the regular airmail route. The plane left San Francisco on 5 October 1922, stopped at Reno, Salt Lake City, Rock Springs and Cheyenne, Wyoming, North Platte and Omaha, Nebraska, Iowa City, Chicago, and Cleveland, and ended the trip at Mineola, New York, on 8 October. Her total flying time was 27 hours, 11 minutes, and covered an estimated 2680 miles. She carried with her souvenirs of several dead aviators: the baby shoes of one, Lincoln Beachey's cuff buttons, and Harold Coffey's goggles. At each stop she addressed the people who gathered, telling them that she was making the flight because she wished to "preserve the memory of these men and many like them who died as martyrs to aviation whether in civil pursuits or in the cause of their country" (author unknown, 1922:1).

When asked by reporters at Mineola to comment on her feelings about her trip she replied (author unknown 1922:1):

> It was a good deal of a rest. Flying is the ideal method of traveling, no cinders, no invitations to buy products advertised on sign boards extending from coast to coast, nothing to disturb the easy sailing through the atmosphere.

Western Air Express was ready to capitalize on the woman in the air. On 17 April 1926, they opened their passenger service between Salt Lake City, Utah, and Los Angeles, along with their airmail run. That same year Maude Campbell became the first woman to fly on Western Air Express from Salt Lake City, Utah. Upon arriving at the airfield she was given a flying suit, a leather helmet, and goggles. As an added safety precaution she was handed a parachute to wear and told to jump if there was any trouble. During the flight in the Douglas M-2 the pilot, Al DeGarmo, passed her notes about their location, speed, and altitude, since the noise of the engine made conversation impossible. At Las Vegas, C.C. Mosley, operations manager of the company, took over the flight. When she landed in Los Angeles Maude was greeted by H.M. Hanshue, president of Western Air Express, and a group of photographers and reporters. Although Maude Campbell was not seeking publicity (she had flown to Los Angeles to visit friends), she became an instant celebrity. Her picture and the story of her flight made the front page of the *Los Angeles Daily News*.

The commercial value of aviation was not only in air travel and transport itself, but in the promotion of goods and services. The Night Aero Advertising Corporation placed an ad in the *New York Times* seeking a woman interested in flying an airplane. They had come up with an advertising scheme of sending an electrically lighted airplane announcing new motion pictures over Times Square at 11 p.m., just as the crowds were leaving the theaters. They were well aware that if the public knew a woman was flying the plane, the advertising would attract more attention. It was a novel idea, but the corporation was unprepared for the response to their ad; 300 women showed up to apply for the job.

Recognizing the persuasive value of women seen using their products, manufacturers featured them

[5] Though Lillian Gatlin did not succeed in having a day set aside specifically for aviators, 11 November was later designated as Armistice Day to honor the memory of all those men who died in the service of their country during World War I.

FIGURE 18.—Women were featured in aviation advertising for everything from planes to goggles. (SI photo 79-14297)

in advertisements for everything from goggles to spark plugs. Manufacturers particularly sought out the endorsements of well-known woman pilots. Kendall Oil published the testimonial letters of Louise Thaden and Phoebe Fairgrave Omlie; Veedol Oil claimed that Thea Rasche used their products; and Elinor Smith appeared in advertisements for Luxor goggles and for Irvin Parachutes.

Parachutes were not commonly used by most pilots. Many thought it "sissy" to fly wearing parachutes; others simply found them cumbersome and uncomfortable. The advertisements that appeared in magazines were geared to convincing pilots and passengers that parachutes could save lives and were not just a means of providing thrills to jumpers and their audiences. In October 1922, Lt. Harold R. Harris's life was saved because he had worn a parachute. As a result, his fellow officers presented him with a watch and a "Caterpillar" certificate. The caterpillar was symbolic of the silk used in early parachutes. From then on, anyone who made an emergency parachute jump could become a member of the Caterpillar Club.

On 28 June 1925, Irene McFarland became the first woman to save herself by the use of a parachute. During a routine aerial exhibition her Thompson chute, which was attached to the bottom of her airplane, failed to detach and open when she jumped. Fortunately, she was wearing an emergency chute and was able to disentangle herself from the Thompson chute and drop safely.

On 1 September 1929, Fay Gillis became the second woman to join the ranks of the Caterpillar Club, and the first woman to jump with a parachute from a disabled aircraft. She was on a test flight of a Curtiss Fledgling with Lt. John Trunk, her flying instructor, when the tail of the plane fell off. Both were able to jump to safety. Fay Gillis, beaming from the pages of aviation magazines with her chute in hand, was a convincing argument for the practical use of parachutes.

Manufacturers also sought endorsements of the aircraft these women flew. These endorsements frequently evolved into full-time paying positions as sales representatives and demonstrators. The women were equally eager to fill these jobs, as they were afforded the opportunity to fly at company expense. They were also confident that they made

FIGURE 19.—Irene McFarland with the parachute that saved her life. (U.S. Air Force Photo)

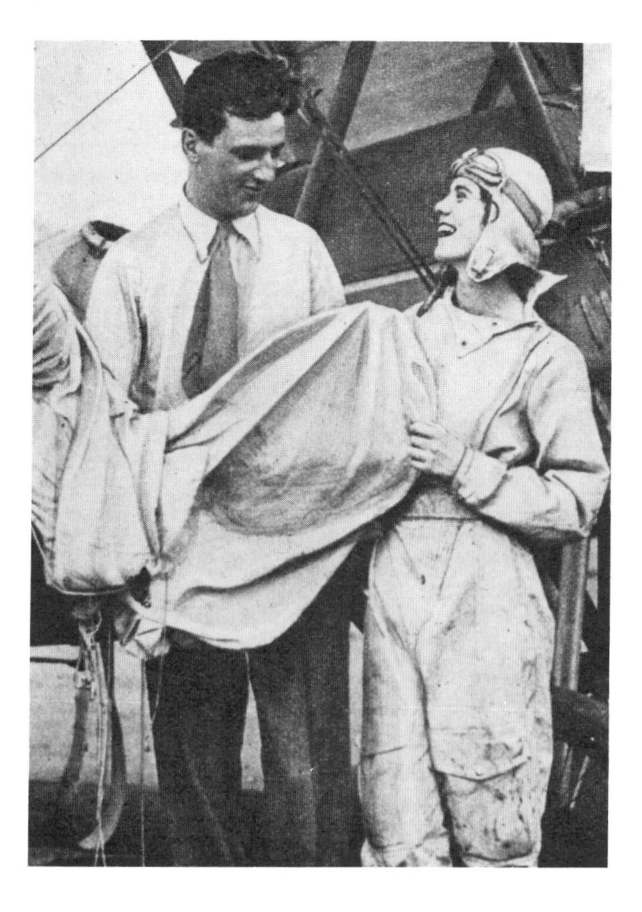

FIGURE 20.—Fay Gillis with Lt. John Trunk as they hold an Irvin parachute. (SI photo 78-2891)

a positive contribution to the selling of aviation to the public in these positions. Evelyn "Bobbi" Trout worked for Golden Eagle, Louise Thaden for Travel Air, Neva Paris and Margery Brown for Curtiss, and Ruth Nichols for Fairchild.

The underlying reason for hiring women to endorse products was simple to understand, and at that time, perhaps a convincing argument: if a woman can use a parachute to bring her to safety, then it is a good idea for everyone flying to use one; if a woman can pilot a plane safely, then anyone can.

There was yet another aspect to this use of women to promote aviation. Selling the idea that flying was safe was of paramount concern to pilots and the industry. In July 1929, Transcontinental Air Transport (TAT) of St. Louis, Missouri, appointed Amelia Earhart assistant to the general traffic manager. She visited cities where airports had been established and delivered speeches promoting the safety and convenience of commercial air transportation to civic and professional associations. In December 1929 she became the company's assistant technical advisor, and she continued to travel and lecture in that capacity. Her principal role, however, was to inspect TAT's facilities and make recommendations for improvements.

Aviation companies sponsored races, record-breaking flights, and other special events featuring

FIGURE 21.—Amelia Earhart (with flowers) was among those who made a promotional flight in one of TAT's Ford Trimotors. Anne and Charles Lindbergh, Edward Bellande, pilot on the flight, and C.S. Wilson are at the right. (SI photo 78-2891)

women pilots. They knew that the attitude of the press would have a considerable impact on the success of these industry-wide efforts to change the image of flying. A sympathetic and supportive press was needed, and women were important in filling that need.

Realizing the power of the written word, women, both pilots and nonpilots, wrote about aviation. They not only extolled the delights of flying, but argued their case for women's right to the sky. In 1924 Margaret Davies wrote a column for *Aero Digest*. She often waxed poetic as she strove to capture the imagination and enthusiasm of the reader and convey the feeling of flight. Katherine Vaughan interviewed woman passengers and reported on what they thought about flying. Ruth Nichols, already a well-known pilot, related her personal flying experiences to the reading public. In 1927 Amelia Earhart joined the staff of *Cosmopolitan* to provide monthly aviation articles.

Some of these women were very outspoken about the rights and obligations of women to fly. Daisy Elizabeth Ball started a career devoted to aviation by working for the staff of *Aircraft* and later became Assistant Secretary of the Aero Club of America. In 1923 she urged that women "take the keenest interest in aviation and educate ourselves to capably fulfill our part in its development" (Ball, 1923:25).

Margery Brown's articles on flying appeared in a variety of magazines in 1929. No one would have thought that Margery, 4 feet, 11 inches in height, would be able to reach the controls of an airplane, but she simply put cushions on the seat and had rudder extensions installed. In her articles she sought to dispel some of the myths about women flyers, sway the opinions of male pilots, and encourage women to take up flying for its character-building qualities. She warned (Brown, 1929:64):

> If you are a woman, and are coming to the flying field seeking stimulation, excitement and flattery, you had better stay away until flying is a little bit safer. If you are thinking that flying will develop character; will teach you to be orderly, well-balanced; will give you an increasingly wider outlook; discipline you, and destroy vanity and pride; enable you to control yourself more and more under all conditions; to think less of yourself and your personal problems, and more of sublimity and everlasting peace that dwell serene in the heavens . . . if you seek these latter qualities, and think on them exclusively, why—FLY!

Lady Mary Heath was a staunch supporter of American aviation. She was a frequent visitor to the United States and during her stays participated in aviation events. She also spent time lobbying in

FIGURE 22.—Margery Brown was an avid pilot, and frequently wrote about flying. Note the signatures on her helmet; she had all the notable pilots she met autograph her helmet. (SI photo 79-6347)

Congress for the cause of women in aviation and pressed the Federation Aeronautique International/National Aeronautics Association (FAI/NAA) for the establishment of women's categories in record flights. She also contributed to the cause by writing for United States aviation and popular magazines, and though her arguments were enthusiastic, unfortunately her facts were not always accurate.

Many female journalists used the press as a forum to promote airmindedness and the development of commercial aviation, but not all advocated that women pursue flying careers. Theodora Marcone was one such writer. She learned to fly in a Dumont Demoiselle in Turin, Italy, in 1910. Though she enjoyed flying as a sport, she did not think women were "physically adapted to undertake aviation as a serious business" (author unknown, 1924:7). Nevertheless her interest in aviation, and particularly its development in the United States, led her to conduct a survey of some prominent businessmen on their use of commercial aviation. Her findings were published in the *Aeronautical Digest* (Marcone, 1924), for which she was a contributing author. She concluded that most businessmen were unaware of the potential advantages commercial aviation could offer them in terms of increased sales and distribution. She further concluded that in order to draw this to the attention of businessmen, the aviation community would have to pull together for the greater good of the industry: to work together to publicize and promote commercial aviation.

Grace W. Smith and Frances Brown, *Air Transportation* staff correspondents, and Alice H. Yost and many other contributing writers reported on the daily developments in aviation. These writers did a creditable job of promoting the industry.

In the mid-twenties the U.S. government began taking a more active interest in the development of civil and commercial aviation. The government had sponsored the early establishment of transcontinental routes for the U.S. Air Mail. In 1918, U.S. Army pilots began flying the mail; they were soon replaced by civilian pilots working for the Post Office Department. In 1923 the Post Office Department began installing beacons along the routes, enabling pilots to fly at night. Airmail routes continued to expand. With the passage of the Kelly Air Mail Act in 1925, contracts to fly the airmail were opened to private companies for competitive bids. Airmail contracts meant guaranteed business; but these contracts were awarded to the lowest bidder. To offset the low bids, some companies began carrying passengers on a regular basis.

During this time Congress had been reluctant to impose any restrictions on flying, whether military or civil. But with the continued increase in civilian air traffic came the inevitable need for regulation. On 20 May 1926, President Calvin Coolidge signed the Air Commerce Act into law. The act charged the Secretary of Commerce "to foster air commerce; designate and establish airways; establish, operate, and maintain aids to air navigation (except airports); arrange for research and development to improve such aids; license pilots; issue airworthiness certificates for aircraft and major aircraft components and investigate accidents" (Briddon and Champie, 1966:1). Government recognition and support, even through regulation, gave respectability to flying. It removed it from a position of novelty to one of stability.

Challenging the Oceans

If aviation profits by the lessons of disaster, as it is preparing to profit, then it may be said that the price, however great, was not too much.

(Author unknown, 1927e)

On 20 and 21 May 1927, Charles A. Lindbergh winged his way across the Atlantic from New York to Paris in 33 hours and 30 minutes, thrilling people around the world. His was not the first transoceanic flight, but it was the longest nonstop solo flight over an ocean. It was a feat, which, until it had been accomplished, had been considered foolhardy, impracticable, and even impossible.

Lindbergh's success was an inspiration to the aviation community, including the women who wanted to be part of the excitement and danger of those early years. They wanted to win laurels like those bestowed on Lindbergh and at the same time to show the American public that aviation was not an exclusively male domain. A transoceanic flight seemed a sure method of attracting such attention, as Lindbergh's flight had proved. These flights were not to be undertaken lightly; many men had died during such attempts. Aware of the risks involved, women were drawn to transoceanic flights not only for the glory involved, but also out of their desire to make a positive contribution to the development of aviation.

James D. Dole, President of the Hawaiian Pineapple Company, hoped to lure Charles Lindbergh into a repeat performance when he announced that he was sponsoring a nonstop flight from Oakland, California, to Honolulu, Hawaii, with a first prize of $25,000 and a second prize of $10,000. Because there had already been two successful nonstop flights to the Hawaiian Islands many assumed it would be an easy race to win. What ensued was a "Mad Hatter" race that left ten dead in its wake. Three entrants died in air accidents before reaching

FIGURE 23.—This Buhl, named the "Miss Doran" for its passenger, was provided by William Molloska for the Dole race. (SI photo 81-3806)

Oakland, five were lost at sea during the race, and two more disappeared during the search for the five already missing.

The race, announced on the heels of Lindbergh's New York-to-Paris success, was scheduled for 12 August 1927, allowing would-be entrants little time for preparation. Initially, three women aspired to make the flight. Pauline Rich of Los Angeles planned to accompany Garland Lincoln; Constance Erwin and her husband, William P. Erwin, planned to enter as copilot and pilot; Mildred Doran planned to fly as John A. (Augy) Pedlar's copilot. Lincoln never actually entered the race, and Constance Erwin was denied permission to enter the race due to a National Aeronautics Association (NAA) rule that prohibited anyone under the age of 21 from participating in any flight under its auspices. But Mildred Doran, a 22-year-old school teacher from Flint, Michigan, was permitted to participate, not as copilot, but as passenger.

Like so many others, Mildred Doran had caught "flying fever" after taking an airplane ride at a local air show. From then on, she spent her spare time at the Flint airport catching rides from whoever was flying. Her special friends were Augy Pedlar and Eyer Sloniger, from whom she gained

FIGURE 24.—Mildren Doran posed for photographers the day before the race. (SI photo 78-3683)

experience handling airplanes, though she never obtained a license.

When she heard about the Dole Race, Mildred decided that she was going to be the first woman to fly to Hawaii. Both Pedlar and Sloniger wanted to enter the race as well, but they knew that only one would be able to make the flight with Mildred. She intended to be the copilot on the flight. They tossed a coin and Pedlar won the toss. William Molloska, a local oil distributor, sponsored their flight. He provided them with a Buhl, which was appropriately christened "Miss Doran." The Buhl was the only biplane entered in the race, and it was hardly the best selection for a long-distance flight over water.

When the crew arrived in Oakland it looked as if they would not be making the flight. Mildred did not have sufficient training to qualify as a navigator, and without one, Pedlar would not be allowed to enter the race. They decided to take a third person, and Lieutenant Silas R. "Cy" Knope joined the crew of the "Miss Doran" as navigator and copilot. There was no official ruling to prevent Mildred from going as a passenger.

Marguerita Jensen, wife of Martin Jensen, who piloted the Breese "Aloha," was also involved in the Dole Race, though not as a competitor. When Martin went to California, she was given power of attorney. She made the arrangements to mortgage his business aircraft so that he could purchase the Breese and cover any additional expenses. Although they won the second place purse of $10,000, the money went to creditors. They made no profit on the venture but they hoped that the national exposure would be good for their business.

Until their divorce Marguerita continued to play a role in Martin's aviation career. Most notable was her participation in an endurance flight on 26 June 1929. This was Martin's third attempt to break the record of 172 hours 32 minutes 1 second. Their plane was a Bellanca monoplane christened "Three Musketeers" and William Ulbrich was flying as copilot. The flight was terminated after 70 hours, 37 minutes, and 58 seconds when high winds and engine problems prevented the refueling ship from taking off.

Marguerita Jensen was like many who entered aviation in the twenties. They were drawn by the romance and excitement of flying, only to find out that there was far more hard work than time for fun. Commercial aviation was fighting its way into existence, but for most pilots it was a tenuous way

FIGURE 25.—Ruth Elder and George Haldeman posed for photographers shortly before their attempt to fly the Atlantic. (SI photo 79-12295)

to make a living. As the novelty wore off, many of these aviation enthusiasts returned to their former, more traditional lifestyles.

Much to the chagrin of the Hawaiian promoters, the contestants insisted on a postponement until 16 August to allow for some additional preparation. Still, most of the entrants, including Mildred, were counting on good luck rather than careful planning to get them to Hawaii. Of the eight planes ready for takeoff, only four got underway. The "Miss Doran" took off at 12:31 p.m., but returned to have misfiring spark plugs replaced. Shortly after 2:00 p.m., they took off again from Oakland Airport and headed for Honolulu. The "Miss Doran" was sighted passing the Farallon Islands at 3:03 p.m. on 16 August 1927, and was never seen again. Mildred Doran was the first woman to attempt to cross an ocean in an airplane, and the first female casualty of such a trip.[6]

[6] The U.S. Navy's search for the missing flyers covered 540,000 square miles between Oakland and Honolulu, utilized 39 vessels, and cost the American people millions of tax dollars.

FIGURE 26.—Ruth Elder suits up for the flight at Roosevelt Field. (SI photo 51326)

The disappearance of Mildred Doran on her flight to Hawaii did not dampen the spirits of Ruth Elder or Frances Grayson. September of 1927 saw both women preparing for transatlantic flights, though it was really too late in the year for such an attempt. The weather would worsen as each day passed, increasing the danger.

Twenty-three-year-old Ruth was living in Lakeland, Florida, when she heard the news of Lindbergh's New York-to-Paris success. Ruth, as did many other women, vowed to be a "Lady Lindy," the first woman to fly across the Atlantic. When she proposed the flight to George Haldeman, her good friend and instructor, he agreed to join her for the flight. With the backing of Florida and West Virginia businessmen, George and Ruth obtained a Stinson Detroiter for the flight, which they christened "American Girl."

There were many who looked on Ruth Elder's Atlantic attempt as a publicity stunt, and in part, it probably was. It certainly would have been more prudent to wait until spring when the weather would be better, but by then public interest might have waned or someone else might have succeeded while she sat waiting. Nevertheless, Ruth was no fool. Her choice of George Haldeman as copilot was quite deliberate, as was the selection of the plane. The Stinson was chosen because of its already proven ability in long-distance flights. Further, they carefully mapped out a flight plan, which, though not the most direct route, would avoid the worst of the Atlantic storms.

With news of the Dole disaster, Ruth's backers suggested she wait a year before attempting the Atlantic, and participate instead in the New York-to-Spokane cross-country race. She was intent, however, on making her attempt before the year was out. When she arrived at Roosevelt Field in September, it looked like the flight would have to be scrubbed. J.J. Lannin, owner of the airfield, did not want her to take off from his field because of the many recent air tragedies. But in the next few days this problem was resolved as Ruth demonstrated her flying ability and received her license from the NAA. George Haldeman would be the pilot and Ruth would accompany him as copilot. The "American Girl" and her crew were ready to go; all that was needed was a break in the weather. On the morning of 11 October 1927 the weather was still poor, and no break was in sight. Nevertheless, they decided that they would take off. The "American Girl" carried 520 gallons of fuel, which would provide an estimated 48 hours of flying time. Based on the fact that Lindbergh had crossed in 21 hours and 40 minutes in poor weather, it was judged that even with a longer route and adverse weather conditions, the "American Girl" had more than enough fuel to reach Europe. But 13 October arrived with no word from the "American Girl;" the flight was now overdue.

The press, who until this time had kept coverage of the proposed flight low keyed (probably because they did not think that this attractive "modern" girl was really serious), hit the front pages of newspapers around the world with expressions of concern and hope for her safety. The *New York Times* (author unknown, 1927b:1) reported that

> everybody in France is eager to see this audacious girl succeed in proving that she is no weak woman. If she does succeed, that lovely American will have a triumph as great as that of Lindbergh.... The daring and self-confidence of the American girl have imbued public opinion with the conviction that she will succeed so that nowhere will be heard repetitions of the pessimistic predictions which sought to discourage flights since the recent series of transatlantic air disasters.

FIGURE 27.—After the crash Ruth was taken aboard a Dutch freighter, the *Barendrecht*. (SI photo 51328)

On 14 October the world was greeted with the news that Ruth Elder and George Haldeman were alive! The plane's oil pressure had dropped to five pounds when they were still far from shore. Fortunately they spotted the Dutch freighter *Barendrecht*. They circled the freighter, brought the "American Girl" down on the water, and were rescued by the freighter's crew. Before the plane could be hoisted to the deck of the ship, it exploded.

Making the rest of her trip to Europe by freighter was hardly what Ruth expected. She had sincerely hoped to fly the "American Girl" herself, but the weather worsened and George did all the flying. Planning their route over well-traveled shipping lanes saved their lives. Ruth was not able to claim that she was the first woman across the Atlantic, but she and George did establish a new over-water record of 2623 statute miles on a Great Circle course.

After disembarking the *Barendrecht* Ruth traveled on to Paris and was fêted for her accomplishment, although the reception was not as grand as it would have been had she landed the "American Girl" there. She then sailed back to the United States where she pursued a career in vaudeville and the movies.

With the failure of her flight came criticism. The flight was condemned as a foolhardy and reckless endeavor that served no useful purpose. Arthur Brisban, noted columnist for the Hearst papers, stated that he hoped "hereafter American girls will stay on the ground or at least do their flying over land."[7] Spokeswomen of the day expressed their opinions, too. Eleanor Roosevelt (author unknown, 1927c:2) termed the flight "very foolish." Dr. Katherine B. Davis, a sociologist, declared (author unknown, 1927c:2) "There is no woman alive today . . . equipped for such a flight," referring to physical ability and practical experience in flying. Probably the most scathing criticism came from Winifred Sackville Stone (author unknown, 1927c:2) when she commented:

> She showed courage but what good did she do? I am much opposed to people who do things of this kind solely to bring their names before the public. This girl came out boldly and said she cared nothing for scientific advancement, but wanted to be the first woman to fly across the Atlantic . . . .
>
> If she had wanted to show that a woman was capable of accomplishing what a man had accomplished, that would have been another matter. If she had wanted to show that a woman could fly alone like Lindbergh, that would have been laudable. But any one can sit in a machine while there is someone to guide it . . . .
>
> This afternoon I am having as my guests at tea a number of high school girls who have won prizes for fast typing. Any one of them, in learning to be a fast, accurate typist, does far more for the community than a dozen . . . Ruth Elders.

While Ruth Elder was at Roosevelt Field in New York, Frances Grayson and her crew were readying a Sikorsky S-36A, "The Dawn," at Old Orchard, Maine. Frances was a well-to-do real estate broker from Forest Hills, Long Island, who also aspired to be "Lady Lindy." Her friend and co-backer, Mabel Ancker, was Danish, so Copenhagen was her destination. Mabel took the steamer to Denmark to await Frances's arrival.

Frances, although reportedly able to fly, was going as a passenger. She had recruited Wilmer Stultz as pilot and Brice Goldsborough as navigator. The Sikorsky Amphibian was selected for the flight on the supposition that if trouble developed, they would be able to land on the ocean and either make the necessary repairs and take off again or wait until help arrived.

Bad weather and mechanical problems with the

[7] Newspaper clipping, Ruth Elder biographical file, Library, National Air and Space Museum.

aircraft found Frances Grayson still at Old Orchard when word of Ruth Elder's successful takeoff reached her. It looked as if she had been beaten until she heard the news of Ruth's disappearance and rescue. Frances (author unknown, 1927a:3) renewed her efforts, stating, "I am going to be the first woman across the Atlantic, and mine the only ship since Lindbergh's to reach its destination. I will prove that woman can compete with man in his own undertakings."

"The Dawn" took off from Old Orchard on 17 October, only to be forced back. The plane was carrying too much gasoline in the nose tanks. The fuel was transferred to five-gallon cans and stowed in the rear of the plane. Again on 23 October Frances and her crew took off, but 500 miles out the weather thickened, and when engine trouble developed, Wilmer Stultz decided to turn back for Old Orchard. Frances was furious, and her anger was compounded when Wilmer refused to make another attempt so late in the year. She decided to return to New York and consult with Igor Sikorsky about the problems they had had with "The Dawn."

Since she was now without a pilot, everyone assumed that Frances would not be able to make another attempt to cross the Atlantic in 1927. She was determined to go, however, and before too long she had enlisted the aid of Oskar Omdahl as pilot. Brice Goldsborough remained as navigator. On 23 December 1927 "The Dawn" took off from Roosevelt Field, New York, at 5:07 p.m., headed for Harbour Grace, Newfoundland. On this first leg of the journey, "The Dawn" carried an extra passenger, Fred Koehler, an engine expert who would stay behind at Harbour Grace. The plane was last sighted of the coast of Cape Cod at 7:10 p.m. Although the Navy and Coast Guard searched the coastal waters until 31 December 1927, no trace was ever found. The disappearance of Frances Grayson ended transatlantic attempts by women for 1927, but the efforts would be renewed in the spring of 1928.

On the other side of the Atlantic, Mabel Boll, an heiress from Rochester, New York, was in Paris declaring that she would be the first woman to fly from Paris to New York. The "Diamond Queen" (a nickname she had earned by wearing so many of the jewels) planned to fly as a passenger, but first she had to find a pilot with a plane. Mabel offered to pay $25,000 to anyone who would make the flight with her, and had broached the subject to several well-known aviators.

FIGURE 28.—Frances Grayson with pilot Wilmer Stultz. He refused to attempt a transatlantic flight so late in the year. (SI photo 79-12315)

After meeting Charles A. Levine, Mabel was sure that he and his pilot, Maurice Drouhin, would take her as passenger on the "Miss Columbia's" return flight to the United States. Levine, an amateur pilot with no license, had crossed the Atlantic earlier in the year with Clarence Chamberlin. He spoke no French; Drouhin, who had been induced to leave the aviation design firm of Farman, spoke no English and knew little about aerial navigation. Besides the inability of these two men to communicate without an interpreter, personality conflicts developed as preparations for the return flight got underway. Levine then decided that "Miss Columbia" could not carry an extra person on such a long flight. Mabel doubled her offer, but Levine, wealthy in his own right, declined. Relations between Levine and Drouhin continued to deteriorate, and Drouhin obtained an injunction to prevent anyone else flying the "Miss Columbia." Finally out of patience, Levine tricked the policeman and mechanic guarding the plane into letting him warm up the plane, and took off for England.

With Levine in London, Mabel announced that she would buy a plane for the flight, stating, "I

FIGURE 29.—Amelia Earhart was a competent pilot, though she made her transatlantic flight as a passenger. (SI photo 81-1431)

intend to be the first woman to fly from Paris to New York, and Drouhin will take me there. I am trying to find a French plane, but I fear I shall not be able to get one before Spring."[8] She soon changed her mind about Drouhin, however, and followed Levine back to London. He had engaged Captain Walter Hincliffe to fly him back to the United States. He still would not let Mabel accompany them, but she did extract a promise that she could use the plane for an attempt to be made from the United States after his return flight.

It was probably just as well for Mabel that none of the pilots she tried to lure into carrying her as a passenger did so after the tragedies of 1927. Having given up hope of finding a plane and pilot in Europe, Mabel boarded a steamer for home. She was still determined to be the first woman to fly across the Atlantic, and once in New York, she planned to renew her efforts to obtain a plane and a pilot for her flight.

Mabel kept Charles Levine to his word about using "Miss Columbia" to fly her to Europe. Before the Atlantic was attempted, however, she accompanied Levine and his latest pilot, Wilmer Stultz, on a flight to Havana, Cuba. With this 1400-mile flight of 14 hours, 25 minutes made on 5 March 1928, Mabel Boll became the first woman to fly nonstop from Mitchell Field, Long Island, to Havana. Upon her return, Mabel began preparing for her transatlantic flight with the understanding that Wilmer Stultz would be her pilot. Unfortunately for Mabel, Wilmer had other plans.

In Boston, Massachusetts, another group was planning to brave the Atlantic, and it was this group that Wilmer Stultz joined. Mrs. Amy Phipps Guest of Pittsburgh, Pennsylvania, wanted to fly across the Atlantic as a passenger. She had always been interested in aviation, and her husband, Frederick E. Guest, had been Secretary of State for Air in Great Britain. She purchased from Commander Richard E. Byrd a Fokker trimotor that he had once intended to use on a South Pole expedition. He had changed his mind, however, in favor of a Ford trimotor. (Ironically, it was also reported that Mabel Boll was negotiating for the purchase of this plane.) Mrs. Guest renamed it the "Friendship." Pressure from family and friends caused her to change her

[8] Newspaper clipping, Mabel Boll biographical file, Library, National Air and Space Museum.

mind about being a transatlantic air passenger, and Amy Guest reluctantly decided to sponsor a suitable, aviation-minded young woman to fly in her stead. Amelia Earhart was that woman.

At the time, Amelia was a social worker at Boston's second oldest settlement house, Denison House, where she taught English to adults and children. When she had any spare time and money she spent them on flying. Amelia's interest in flying was first aroused in 1918 as she stood and watched pilots training at the military airfield near Spadina Military Hospital in Toronto, Canada, where she worked as a nurse's aid. However, it wasn't until 1920 that Amelia had her first ride in an airplane. After that ride, she knew that she would become a pilot. Her first instructor was Anita "Neta" Snook. When Neta "retired" from flying in 1922 to raise a family, her friend, John "Monte" Montijo, continued Amelia's instruction. Amelia soloed that same year. She had already accumulated 500 hours of flying time when she was approached with Mrs. Guest's proposal.

The members of the "Friendship's" crew would be Wilmer Stultz, pilot; Louis Gordon, copilot; and Amelia Earhart, passenger and recorder for the flight, with Lou Gower as the standby pilot. There was no announcement to the press of the proposed flight. From all appearances, Richard Byrd was preparing the "Friendship" for a flight to the South Pole, even though he was already having tests made on the Ford. Mabel Boll was unperturbed by Stultz's frequent trips to Boston to test the "Friendship." A number of modifications had been made to the craft. Pontoons had replaced wheels, extra fuel tanks had been fitted, and new instruments and radio equipment had been installed.

The plane was ready by the end of May but there was insufficient wind on Boston Harbor for the "Friendship," heavily laden with extra fuel and equipment, to take off. Two unsuccessful attempts were made to get underway, and at last on 3 June 1928 at 6:30 a.m., there was enough of a breeze for takeoff. The "Friendship" was on her way with Amelia Earhart aboard. On the first leg of the journey they were to reach Trepassey Bay, Newfoundland. As they flew north from Boston the fog thickened and at Halifax, Nova Scotia, Stultz brought the plane down. He and Gordon went ashore to check the weather reports while Amelia waited on the plane. Despite reports of rain and clouds all the way to Trepassey, they decided to continue their flight. Soon, however, they turned back for Halifax.

When Mabel Boll heard that Wilmer Stultz was piloting another plane with another woman as passenger, she sobbed to a reporter (author unknown, 1928a:3), "now he has gone and taken off with this other woman and I was sure he would fly with me. I depended on him." It looked as if all hope for Mabel's flight was ended. On 4 June, the "Friendship" left Halifax for Trepassey. The fog was already forming when they came in for their landing. The weather continued to deteriorate during the night and kept them grounded for several days.

Mabel was elated on 5 June to hear that her "rival," Amelia, was fog-bound in Newfoundland. It gave her yet another chance to make good her declaration to be the first woman to cross the Atlantic by air. She had the "Miss Columbia" and Captain Oliver Boutiller had agreed to be her pilot. He had tested the "Miss Columbia" while Wilmer Stultz had been in Boston testing the "Friendship." Arthur Argles agreed to accompany them as navigator, but they could not leave Roosevelt Field until the fog lifted and the rain abated. Finally on 12 June 1928, Mabel was off on the first leg of her journey, which would take her to Harbour Grace, Newfoundland. The press had been awaiting her arrival and wanted to know all about her "race" with Amelia. She denied that she was racing, stating (author unknown, 1928c:2), "I would like to be the first woman to cross the Atlantic, but if the fates are against us I am satisfied to be the second. I wish Miss Earhart and her companions the best of luck." Shortly thereafter she sent Amelia a telegram "inviting" her down to Harbour Grace where the water was smoother, so that they could take off together in a friendly competition.

The winds at Trepassey made the water too rough to even allow the loading of the gasoline in the "Friendship." But on 17 June the weather cleared and the water on the bay was smooth enough for takeoff. At 9:15 A.M., Amelia and the "Friendship" were on their way across the Atlantic. The good weather did not last for long, and they soon encountered dense fog, rain, and strong winds.

Caught at Harbour Grace, Mabel adopted a wait-and-see attitude. There was still a slim chance that she would be the first woman to fly the Atlantic. Instead of taking off immediately, the "Miss Columbia's" crew decided to see if the "Friendship" was really on its way and not just on a test flight. Such a delay could mean losing the "race," but they

FIGURE 30.—Amelia Earhart, flanked by the pilot and co-pilot of the "Friendship," is welcomed by a committee to Burry Port, Wales. (SI photo A43,0333-C)

reasoned that if the "Friendship" chose the Azores route, they could still attempt a direct flight the next day and have a good chance of landing first. And, of course, there was always the possibility that the "Friendship" would be lost at sea, in which case the crew of the "Miss Columbia" would be able to take its time in making an Atlantic flight.

But 20 hours and 40 minutes later Wilmer Stultz brought the "Friendship" down at Burry Port, Wales, and Amelia Earhart became the first woman to cross the Atlantic in an airplane. Amelia was hailed as a heroine, even though she quickly pointed out that the flight only succeeded through the skills of Stultz and Gordon. After a round of teas, lectures, and social engagements, Amelia returned home by steamer.

Back in the United States, Mabel claimed that she had not received the same weather reports as Amelia. The Weather Bureau quickly affirmed that she had not, but pointed out that the reports had been available for Amelia's flight because the "Friendship's" backers had requested and paid for them. They were not available just as a matter of course. Mabel Boll soon recovered from her disappointment and was off again to Europe to see Charles Levine, still trying to persuade him to fly her across the Atlantic.

And what was the point of these transoceanic attempts during 1927 and 1928? Such flights were decried as foolhardly and too dangerous for women. The women involved were criticized as being publicity seekers and for not contributing anything to the science of aviation. Yet they still tried, knowing full well what chances they were taking. Then finally on 18 June 1928, Amelia Earhart had flown across the Atlantic—not as a pilot, as she would have liked, but as a passenger. What did such a flight accomplish? Perhaps a *New York Times* editorial summed it up best (author unknown, 1928b:24):

> What trans-Atlantic aviation needs if it is ever to reach the commercial stage . . . [is] something in the nature of a series of carefully engineered experiments. . . . Because of the circumstances under which she flys Miss Earhart's must be regarded as the first step in an engineering investigation out of which will emerge the practical transatlantic passenger carrying flier of the future.

FIGURE 31.—Clara Adams crossed the Atlantic on the Graf Zeppelin in 1928. (Courtesy of Marjorie Stinson Collection, Library of Congress)

As if to emphasize this point, there was one more epic transatlantic crossing in 1928. This time the Graf Zeppelin would carry eight paying passengers from Lakehurst, New Jersey, to Germany. It was a time when there was much debate as to whether an aircraft large and powerful enough to carry a payload of passengers and cargo over long distances could ever be built. The airship seemed to be the most obvious and economical solution to this debate.

Among the names on the Graf Zeppelin's passenger list was that of Clara Adams, a veteran passenger of flying machines. Clara had made her first flight in a Thomas airboat in 1914. In 1924 she was in Germany when the Zeppelin ZR-3 "Los Angeles" was making test flights prior to its transatlantic flight. She met Dr. Hugo Eckner, designer of the craft, and went on one of the test flights. Clara was not to be allowed to make the crossing in the ZR-3, but Dr. Eckner promised that he would take her on the first commercial flight of one of his airships. Thus on 29 October 1928, Clara Adams became the only woman commercial passenger of the Graf Zeppelin on its first commercial transatlantic flight. (Lady Grace Drummond Hay was listed as a passenger on the Graf Zeppelin's inaugural flight from Germany to the United States, and was thus the first woman to travel on that airship.)

Setting Records

Records are made to be broken, and I only wish that more girls could get good ships and keep setting new marks all the time. It has long been my theory that if women could set up some records, in many cases duplicating the men's, the general public would have more confidence in aviation.

(Ruth Nichols in Buffington, 1974:10)

In their attempts to cross the oceans by air women had shown that they, too, had the courage, daring, and persistence to pursue their aviation goals. Still, they had not been able to demonstrate their skills. Although several had had flight training (Doran, Elder, and Earhart), not one woman had taken the controls during her flight.

More and more women were actively interested in aviation. Though by 1927 many male pilots were still reluctant to take on female students, a growing number of women were enrolling in flying schools across the country; Ila Fox was one such woman. After receiving her first few salary checks as recreational director of the Lend-A-Hand Club, she went to Cram Field in Davenport, Iowa, and enrolled in the flight school there. She attended classes in aeronautics and meteorology as well as working with the mechanics on the airplanes. She took her flight instruction in a Travel Air biplane. As flying time was expensive, Ila made an arrangement with the Curtiss Flying Service in Moline, Illinois. On Sundays and holidays she came to the airfield attired in her most alluring flying togs (black leather riding boots, tight pants, white leather flying jacket with white helmet and scarf). When a crowd had gathered she and a flying instructor took off; after the demonstration she answered questions about flying and at the same time promoted the flying service. On 1 September 1929, Ila became Iowa's first licensed woman pilot.

Most women who flew were, admittedly, only interested in flying for recreation. Many probably thought that flying an airplane would be just like driving an automobile. Although driving in the twenties was still no simple procedure, it did not require nearly as much skill and dexterity as flying. Many discovered flying through passenger services and went on to take instruction. Some "washed

FIGURE 32.—Ila Fox transported passengers and made demonstration flights for Curtiss. (SI photo 81-12625)

FIGURE 33.—Both men and women operated passenger flights and provided flying lessons, operating either from established airfields or any level plot of ground. (SI photo 79-865)

FIGURE 34.—Anita "Neta" Snook, one of the first women to give flight instruction, with her plane and her dog Camber. (SI photo 79-6733)

out'' during flight training, or after discovering that flying was not as simple as it appeared, dropped it. Many other women learned just enough to enable them to get around, but did not apply for licenses. Yet a number of women viewed aviation as a wide open field where they could make their mark.

It was not really unusual for a woman pilot to go into business for herself. As with any pilot, she could be expected to take on students. After World War I, Anita Snook was probably one of the first women to give flight instruction. She set up in business at Kinner Airport in Los Angeles. She, along with such pilots as Elinor Smith and Evelyn ''Bobbi'' Trout, was also involved in carrying passengers for hire, not only at the ever-present airshows, but on point-to-point flights.

Women were also involved in the establishment and operation of airfields across the country. Phoebe and Vernon Omlie ran an air service based in Memphis, Tennessee. Cecelia and Tom Kenny, who learned to fly together, opened their own airfield, the Kenny Flying Service, near Millersport, New York. They ran a flying school, provided charter flights, and sold and serviced airplanes. Esther and Earl Vance opened the Vance Airport just north of Great Falls, Montana. Esther was the business manager and treasurer for the air service they operated. She was also the first woman in Montana to hold a commercial license.

Yet, by the joint nature of these business enterprises, women did not attract much national attention. If women wanted to dramatically demonstrate their flying skills they would have to pursue another route, a route already established by men—setting and breaking records. Men had been setting and breaking speed, distance, and altitude records for years, and their achievements had been duly attested and recorded by the FAI, and the NAA, the U.S. component of the international monitoring agency.

When women began flying to set records, they lacked a history and experience in this kind of flying, and their first flights fell far short of records established by men. As the FAI/NAA had no separate category for women, their first efforts went unrecorded and unacknowledged by this regulatory body. Determined that their records be recognized, women persisted in requesting that their achievements be monitored and recorded by the FAI.

It was, perhaps, an endurance flight by Viola Gentry in 1928 that served as the impetus for the series of record challenges that would follow. Viola decided to take up flying after seeing famous stunt pilot Omer Locklear land his plane on top of a hotel in 1919. She did not, however, get her license until 1925. Her goal for the flight was to remain aloft for thirteen hours, thirteen minutes, thirteen seconds. She firmly believed that thirteen was her lucky number.

On 20 December 1928, at 5:44 A.M. she took off from Curtiss Field, Long Island, in a Travel Air biplane powered by a 125-horsepower Siemens-Halske radial air-cooled engine. The plane had been lent to her by Grace Lyon, a strong supporter of women and aviation. Harry L. Booth was on hand to supervise the flight as representative of the NAA, and before she took off, he checked and sealed the barograph that would officially record her flight.

Although Viola was prepared to face a long, cold, lonely flight, the weather was fair at takeoff. For flying in the open cockpit she had donned two flying suits, woolen head protection, fur-lined helmet, gloves, boots, and a parachute. Shortly after takeoff the fog rolled in and a chill rain began to fall, icing up the plane. The weather continued to deteriorate, and after being in the air for 8 hours, 6 minutes, and 37 seconds, Viola landed.

FIGURE 35.—Viola Gentry, suited up for her attempt to set a women's endurance record. A Very pistol is stuffed into her parachute harness. (Courtesy of Viola Gentry)

Since her flight did not exceed endurance records previously set by men, it was not recognized as a record flight by the FAI. It was, however, the first woman's attempt to be monitored and recorded by the FAI/NAA, and in this sense stands as the first official women's endurance flight.

Two weeks later on 2 January 1929, Viola's "record" was broken by Evelyn "Bobbi" Trout with a flight of 12 hours, 11 minutes in a Golden Eagle over Van Nuys Airport, California. This was also the longest night-flight by a woman to that date. Then on 31 January Elinor Smith remained aloft for 13 hours and 16 minutes in an open cockpit Bird biplane over Mitchell Field, Long Island. Elinor had first gained notoriety when she flew under all four of the bridges spanning New York's East River. At the age of 15 she was the first person to do this successfully. For this potentially dangerous stunt

FIGURE 36.—Bobbi Trout capturing the endurance record over Van Nuys Airport. (Courtesy of Bobbi Trout)

FIGURE 38.—Bobbi Trout receives a send-off for her endurance flight from Elizabeth McQueen, a strong advocate of women's flying. (Courtesy of Bobbi Trout)

FIGURE 37.—Elinor Smith set a new women's endurance record on 31 January 1929. (Courtesy of Elinor Smith)

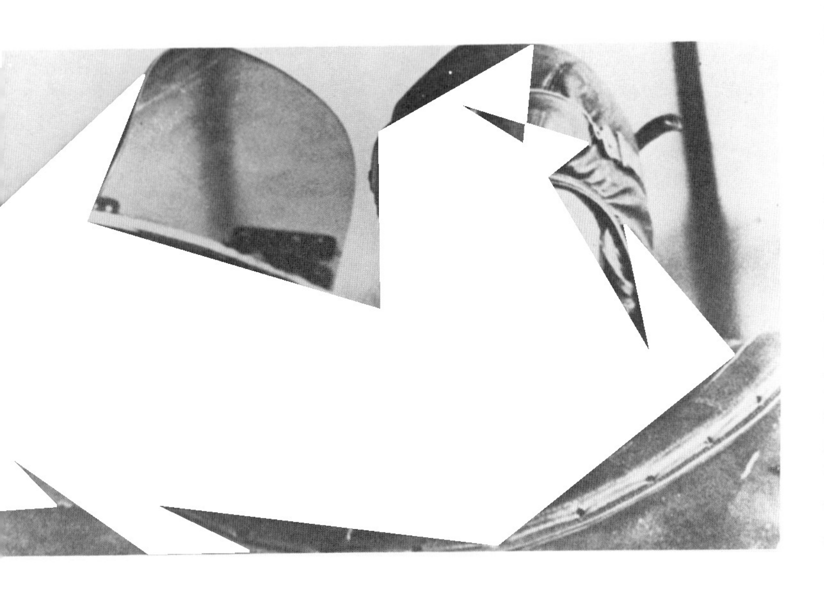

she received a short suspension of her flying license. Now, two years later she was seriously competing in establishing and advancing women's flying records. Bobbi Trout recaptured the record on 10–11 February with 17 hours, 5 minutes, again in the Golden Eagle over Mines Field, Los Angeles. A month later, on 16–17 March, Louise Thaden made her bid for the endurance record from Oakland Municipal Airport, California, in a Travel Air, and succeeded with a flight of 22 hours, 3 minutes. And on 23–24 April, Elinor Smith reclaimed the endurance record with a flight of 26 hours, 21 minutes over Roosevelt Field, New York. The Bellanca CH monoplane flown by Elinor for this record was the largest plane utilized for such a purpose to that date. One might have expected that such a large craft would have been difficult for a 17-year-old girl to handle, but she flew it quite well. Each succeeding flight had been longer not by minutes, but by hours, but women still had a long way to go. The endurance record held by a man was around 60 hours.

FIGURE 39.—Louise Thaden with NAA officials Charles S. Nagel (left) and Dillard Hamilton (right) before her solo endurance flight in March 1929. (Courtesy of Louise Thaden)

FIGURE 40.—An extra gas tank and an extra oil tank were installed in front of Louise Thaden's cockpit for her flight on 16–17 March. (Courtesy of Louise Thaden)

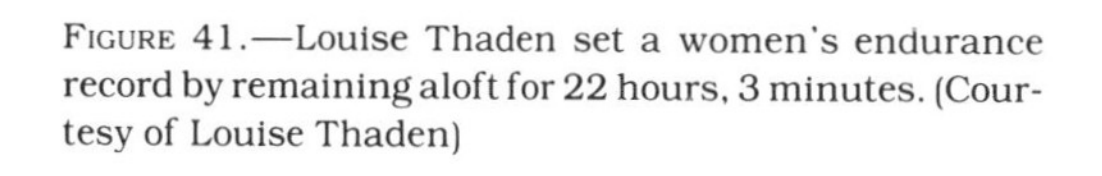

FIGURE 41.—Louise Thaden set a women's endurance record by remaining aloft for 22 hours, 3 minutes. (Courtesy of Louise Thaden)

FIGURE 42.—Elinor Smith stands next to the Bellanca CH she flew for her second endurance record. (Courtesy of Elinor Smith)

FIGURE 43.—Charles S. Nagel, NAA observer for the FAI, installed the barograph that would record Louise Thaden's altitude during the first officially recorded women's altitude flight. (Courtesy of Louise Thaden)

Even before many women were breaking endurance records, others were testing themselves and their planes at ever higher altitudes. As early as 1911, Matilde Moisant reached an altitude of 2500 feet, an impressive feat for the state-of-the-art and a female pilot. Others slowly followed. In July 1913, Alys McKey Bryant reached an altitude of 2900 feet. Three years later, Ruth Law climbed to 11,200 feet, and then in 1917 broke her own record by attaining an altitude of 14,000 feet. Not until 1923 did another woman make an attempt to climb higher; Bertha Horchum attained an altitude of 15,300 feet.

These were unofficial records, and in 1928 did not exceed those set by men. But they were marks at which women could look back as they set out to establish new records. On 7 December 1928, Louise Thaden, piloting a Travel Air, climbed to a height of 20,260 feet, the first officially recorded woman's altitude flight. This was not bettered until 28 May 1929, when Marvel Crosson attained an altitude of 24,000 feet in a Ryan Brougham over Los Angeles, California. Later that year Bobbi Trout in the Golden Eagle set an altitude record of 15,200 feet

FIGURE 44.—Marvel Crosson in the Ryan Brougham B-3 in which she broke Louise Thaden's altitude record. (SI 79-636)

FIGURE 45.—Elinor Smith (left) and Bobbi Trout (right) teamed up for an aerial refueling flight. (Courtesy of Bobbi Trout)

FIGURE 46.—Ruth Alexander set a new light plane endurance record in November 1929. (SI photo 79-7179)

for light planes.

Women had competed to stay up longer and climb higher, and now Louise Thaden intended to fly faster than any woman had done before. On 13 April 1929, she attained a speed of 156 miles per hour in a Travel Air Speedwing.

Since the FAI/NAA had no separate classification for women, these pilots did not receive the recognition they deserved for their flights. Although their flights were now duly recorded, no special significance was accorded to them. These women were determined, however, that their accomplishments would be acknowledged, and their persistence won out. At first women's records were listed under Miscellaneous Air Performances. Finally, at the FAI Conference held in Copenhagen on 19–22 June 1929, the need for a separate category for recording women's record flights was officially acknowledged, and the system was established. From that date, women's records were no longer compared to men's. Now, women's achievements could be seen in a perspective, and they would be judged on the basis of the progress they were making. It was some time before women developed the skills and gained access to the aircraft that enabled them to catch up with and surpass records set by men.

The decision of the FAI Conference gave a new impetus to women's pursuit of record marks. The solo endurance record (non-refueling) was not successfully challenged again by a U.S. woman in 1929. Bobbi Trout and Elinor Smith cooperated on 27–29 November on an in-air refueling flight over Los Angeles that lasted 42 hours, 5 minutes. Ruth Alexander of San Diego captured the light-plane altitude record from Bobbi Trout when she took her Great Lakes airplane to a height of 15,719 feet on 18 November 1929. She had worked in a beauty parlor to earn the money to pay for her flying lessons, and felt that all her hard work had paid off when she set this record. To close out the year, on 9 December Amelia Earhart set a new speed record of 184.17 miles per hour in a Lockheed Vega.

Competitive Racing

> To us the successful completion of the Derby was of more import than life or death. Airplane and engine construction had advanced remarkably near the end of 1929. Scheduled air transportation was beginning to be a source of worry to the railroad. Nonetheless a pitiful minority were riding air lines. Commercial training schools needed more students. The public was skeptical of airplanes and air travel. We women of the Derby were out to prove that flying was safe; to sell aviation to the layman.
>
> (Thaden, 1938:77)

Although women had been involved in many aspects of flying—exhibition, sport, transport, sales, and record setting—one area remained to be tackled—racing. Women wanted to participate in competitive flying for the same reasons men did. This was a very dramatic and visible way for pilots to demonstrate their abilities, and of course, the cash prizes were an extra incentive.

On 13 August 1929, one of the most exciting aviation events since Lindbergh's New York-to-Paris flight began. The National Women's Air Derby, a cross-country race, began that day at Santa Monica, California. The race attracted women from all over the United States, and even took on an international flavor with the participation of a German and an Australian.

Of the 70 women who held U.S. Department of Commerce licenses, only 40 met the minimum requirements for the race: each entrant was required to have flown 100 hours solo, 25 of which had to be cross-country flights of more than 40 miles from the starting point. Besides being licensed by the Department of Commerce, each entrant had to hold a FAI license and an annual sporting license that was issued by the contest committee of the NAA. Each entrant's plane had to carry an Approved Type Certificate and hold a license issued by the Department of Commerce.

Initially the rules allowed for pilots to carry mechanics with them, but this was changed when Hollywood came forth with a list of actresses who wanted to participate. It was evident that these actresses with their mechanic/pilots were no asset to the race. The rules were changed to state that each entrant would fly alone. This did not mean that the women would not have their own mechanics available, however. Wiley Post and Carl Lienesch flew escort, and other mechanics followed the race to be on hand to service planes along the way.

A number of safety precautions were also included in the rules governing the race. Each pilot was required to wear a parachute. Although they

FIGURE 47.—Ruth Elder's Atlantic flight helped further her movie career. She acted in a number of films, including Hoot Gibson's "Winged Horseman." (SI photo 79-12296)

were heavy, uncomfortable to wear, and hampered movement, everyone faithfully wore her parachute throughout the race. Each pilot was also required to carry a gallon of water and a three-day supply of food. Much of the terrain over which these pilots flew was hazardous. With the limited navigational aids at their disposal, getting lost was a distinct possibility. As mechanical failure was not unusual, a supply of emergency rations might well be needed.

Two classes of planes were eligible for the Derby. The CW (light-plane) Class were those powered by an engine or engines with a total piston displacement of more than 275 but not more than 510 cubic inches. The DW (heavy-plane) Class aircraft were those powered by an engine or engines with a piston

FIGURE 48.—Amelia Earhart planned to fly in the Derby as training for a second transatlantic flight, this time as pilot. (SI photo 79-6354)

FIGURE 49.—Evelyn "Bobbi" Trout crouches next to the Golden Eagle Chief provided by R.C. Bone. The National Exchange Clubs (note the emblem on the fuselage) were the hosts and sponsors of the race. (Courtesy of H. Glenn Buffington)

FIGURE 50.—Marvel Crosson next to the Travel Air she flew in the Derby. (Courtesy of H. Glenn Buffington)

FIGURE 51.—Thea Rasche, Derby entrant from Germany, with her Gypsy Moth in the background. (Courtesy of H. Glenn Buffington)

FIGURE 52.—Ruth Nichols stands next to her Rearwin. (Courtesy of H. Glenn Buffington)

displacement of more than 510 but not more than 800 cubic inches.

Twenty women entered the Derby. They faced eight grueling days of flying. The route to Cleveland, Ohio, took them over deserts, mountains, and plains. They navigated by dead reckoning, aided only by temperamental compasses and roadmaps. They encountered route changes, sabotage, and death.

Several of the entrants were well known. There to catch the reporter's eye was Ruth Elder. After her Atlantic attempt she had been accused of using the stunt to further her acting career. The flight *had* put her in the limelight, but she had not abandoned flying. When she qualified for the FAI license that was necessary for participation in the Derby, Ruth had logged 300 hours of flying time.

Amelia Earhart was also entered. Since her Atlantic flight she had been in great demand as a lecturer, speaking not only on her "Friendship" experience, but on her views of women's future in aviation. She was also writing for magazines. In August 1928, she had become an associate editor for *Cosmopolitan*. Amelia entered the Derby not only because it was the first such challenge offered to women, but also because her Atlantic flight made her realize how much skill she still needed to acquire for an Atlantic solo. The Derby was one of the trials she had set for herself in preparation for such a flight.

Evelyn "Bobbi" Trout had taken her first airplane ride on 27 December 1922 when she was sixteen. It was not until January 1928, after saving enough money and being able to take the time off from managing her father's filling station, that she had her first flying lesson. She soloed on 31 April 1928, and spent every spare moment building up flight time toward her private pilot's license. When R.O. Bone, manufacturer of the Golden Eagle, offered her a job demonstrating his plane, she accepted. It was in his plane that she set so many records. When the Derby was announced, Bone provided Bobbi with the Golden Eagle Chief to use in the race.

At the age of 25, Marvel Crosson was looking forward to a career in aviation. She had taken up flying in 1922 while living in San Diego. Her brother Joe taught her to fly, and she soloed in 1923. She soon followed Joe to Alaska where she flew commercial aircraft until returning to San Diego. For the Derby she flew a Travel Air provided by Walter Beech.

Thea Rasche had taken up flying in 1924 in

FIGURE 53.—Jessie Maude Keith-Miller did not place in the Derby, but she did claim a $500 purse in closed-course events at Cleveland. (Courtesy of H. Glenn Buffington)

FIGURE 54.—Phoebe Omlie next to her Monocoupe, a plane she often demonstrated for its manufacturer. (Courtesy of H. Glenn Buffington)

FIGURE 55.—Louise Thaden prepares for takeoff at Santa Monica, California. (Courtesy of H. Glenn Buffington)

Münster, Westphalia, Germany. In 1925 she qualified for both her pilot's license and an aerobatic license, becoming Germany's first female stunt flyer. By 1929 she was already well known in the United States, where she had been flying in exhibitions to promote airmindedness. When plans for the Derby were announced, the Moth Aircraft Corporation of Lowell, Massachusetts, asked Thea to fly a de Havilland Gypsy-Moth for them; she agreed. The promised plane was not ready in time for the race, however, so another older Gypsy-Moth was located for her in Los Angeles. Thea had to take off without having had an opportunity to test the plane.

Ruth Nichols was one of the old-timers; she had been flying since 1922. Now she was involved in promoting airplanes as a practical means of transportation. In 1928 after a non-stop New York-to-Miami flight in a Fairchild monoplane, Ruth joined the sales department of Fairchild Airplane and Engine Company. She was also involved in Aviation Country Clubs. To publicize the clubs Ruth flew in a 12,000-mile Sportsman Air Tour, thus becoming the first woman to land in all the 48 states.

Jessie Maude "Chubbie" Keith-Miller embarked on an aviation career quite by chance. While in London she met Bill Lancaster, who was planning to fly from England to Australia. In return for raising half the money, Chubbie accompanied Bill on the flight. They took off from Croydon on 14 October 1927, in an Avro Avian on a reliability flight. Though they were not the winners when they completed their flight on 19 March 1928, Chubbie was the first woman to fly to Australia. When she took off with Bill, she did not know how to fly, but by the time she reached Australia she was a competent pilot. The promise of easy money and aviation opportunities lured the two to the United States. When Chubbie heard about the Derby and that there would be a large purse for the winner, she decided to enter. After obtaining a license, she approached Lawrence Bell and convinced him to provide her with a Fleet biplane.

Phoebe Fairgrave Omlie was perhaps the best prepared for the rigors of the race. When the Mississippi River went on the rampage in 1927, she and her husband flew relief to flood victims until the river began to recede. Then in 1928 she became the first woman to complete the National Air Reliability Tour, a "race" that covered 5304 miles, 32 cities, 13 states, and lasted almost a month. She also spent much of her time flying throughout the United States, Canada, and South America as special assistant to Don Luscomb, president of Mono Aircraft. She often demonstrated his Monocoupe, the same plane she flew in the Derby.

Louise Thaden was working for the J.H. Turner Coal Co. in 1926, but she spent so much time visiting the Travel Air Factory that Turner, aware of her interest, introduced her to his friend Walter Beech, who owned Travel Air. Beech offered her a job with his Pacific Coast distributor, and she accepted. As part of her salary, Louise received flying lessons. In California, Louise began making record flights in Beech Travel Airs. In December 1928 she set a women's altitude record of 20,260 feet. In March 1929, she flew for 22 hours, 3 minutes, 28 seconds for a women's endurance record, and then in April 1929, she set a women's speed record of 156 miles per hour. Louise was the first woman to hold all three records simultaneously.

Eleven of the entrants were relative newcomers to flying. Opal Kunz's entry in the Derby may have been a surprise to some of her friends. A society leader and wife of George Kunz, vice-president of Tiffany Co., she had only begun flying in March 1929. At first glance it might have appeared that her enthusiasm was just faddish, but she was quite serious about flying. She felt very strongly that women should fly not only for sport but also as a career, that American women should be able to compete with European women in flying, and that learning to fly should be part of the national preparedness for both men and women. Opal had purchased her own plane for the race, a Travel Air equipped with a 300-horsepower Wright Whirlwind engine but it was ruled ineligible. Fortunately, she was able to acquire a Wright J-5 equipped Travel Air.

Vera Dawn Walker started flying in the fall of 1928 at the Standard Flying School in Los Angeles. Her short stature (4 feet 11 inches) in no way diminished her desire to fly. She simply propped herself up on pillows so that she could reach the rudder pedals and commenced flying. On 1 January 1929, she received her Department of Commerce license, and in July her FAI license. Though she had little time to devote to flying (she worked two jobs—selling real estate and as an extra in the movies),

FIGURE 56.—Opal Kunz in her Travel Air. To the right is Elizabeth McQueen, a strong supporter of this event. (Courtesy of H. Glenn Buffington)

FIGURE 57.—Vera Dawn Walker of Los Angeles poses next to her Curtiss Robin. (Courtesy of H. Glenn Buffington)

FIGURE 58.—Margaret Perry, a newcomer to flying, was already a commercial pilot. She flew a Spartan. (Courtesy of H. Glenn Buffington)

she managed to accumulate the hours necessary to qualify for the Derby.

Margaret Perry began her career in sport and commercial aviation in 1929. After starting her flying lessons at Curtiss Field on Long Island, she continued them at the California Aerial Transport Field in Los Angeles. Margaret was quite serious about her flying as a commercial pilot. She had qualified for her transport license and had purchased her own plane.

In 1929 flying was still considered a male pursuit, and women who took it up either for recreation or as a career were often labeled "tomboys." The only Derby entrant who really fitted that description was Florence Lowe "Pancho" Barnes. It was no surprise that Pancho took up flying. Her family was already active in aviation; her grandfather was Thaddeus S.C. Lowe, the pioneer aeronaut. When she qualified for her license in 1928 she was only following in his footsteps.

Neva Paris qualified for her private license on 31 December 1928. Although a native of Kansas City, Missouri, her flying career had taken her across the country to New York. There she had a position in the demonstration and sales department of Curtiss-Wright at Valley Stream, Long Island.

Mary E. von Mach received her private license on 15 November 1928. Mary flew for the sheer fun of flying, in contrast with many of the other entrants, for whom flying was a career. When she qualified

FIGURE 59.—Florence "Pancho" Barnes posed with a group of fellow pilots and sponsors at Santa Monica. (Courtesy of Bobbi Trout)

for her FAI license on 9 August 1929, she had just enough flying time logged to meet the Derby requirement of 100 hours. Mary was also one of two licensed woman pilots in Michigan.

Mary Haizlip had her first opportunity to fly after meeting her husband-to-be, James Haizlip. He had returned from France after World War I as a senior pilot and was running a flying school while attending the University of Oklahoma. She took flying lessons from her future husband, and when she qualified for her FAI license she had already logged 230 flying hours.

Edith Foltz had given up a musical career to join her husband in the field of aviation. In early 1928 they bought an airplane and hired a pilot to teach them to fly. Together they barnstormed across the northwestern United States. In August 1928, Edith became the first woman in Oregon to receive her pilot's license.

Blanche Noyes was pursuing a career in the theater and the movies when she met Dewey Noyes and decided that she would rather fly with him than act. Dewey, a pilot for the United States Airmail, taught her, and Blanche soloed after only 3 hours, 45 minutes of dual instruction. She became Ohio's first licensed woman pilot.

Gladys Berry married James Lloyd O'Donnell in 1921, and it was through her husband's interest in aviation that she began flying. Under his coaching she finally got her private license on 17 May 1929. Then on 5 August 1929, having accumulated 122.5 hours of flying time, she qualified for her FAI li-

FIGURE 60.—Neva Paris in front of the Curtiss Robin she flew in the Derby. (SI photo 79-9684)

FIGURE 61.—Mary von Mach in the cockpit of her Travel Air, the "Mary Ann II." (Courtesy of H. Glenn Buffington)

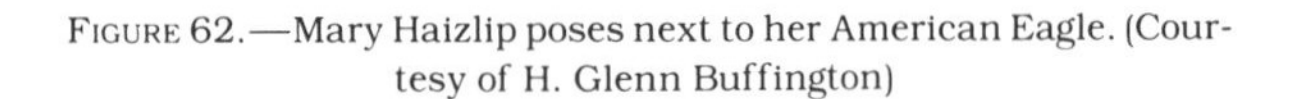

FIGURE 62.—Mary Haizlip poses next to her American Eagle. (Courtesy of H. Glenn Buffington)

FIGURE 63.—Edith Foltz stands by her Alexander Eaglerock Bullet. The roadmaps in her hand were among the few navigational aids available to pilots. In the air there was little opportunity to read maps so entrants studied routes in any spare moments they had. (Courtesy of H. Glenn Buffington)

FIGURE 64.—Blanche Noyes in front of her Travel Air, the "Miss Cleveland," holds an armful of roses from her home town supporters. (Courtesy of H. Glenn Buffington)

FIGURE 65.—Gladys O'Donnell with her husband Lloyd at Douglas, Arizona. (SI photo 79-5996)

FIGURE 66.—Claire Mae Fahy, who was forced to drop out early in the race, points to wires that she claimed had been eaten through by acid. (Courtesy of H. Glenn Buffington)

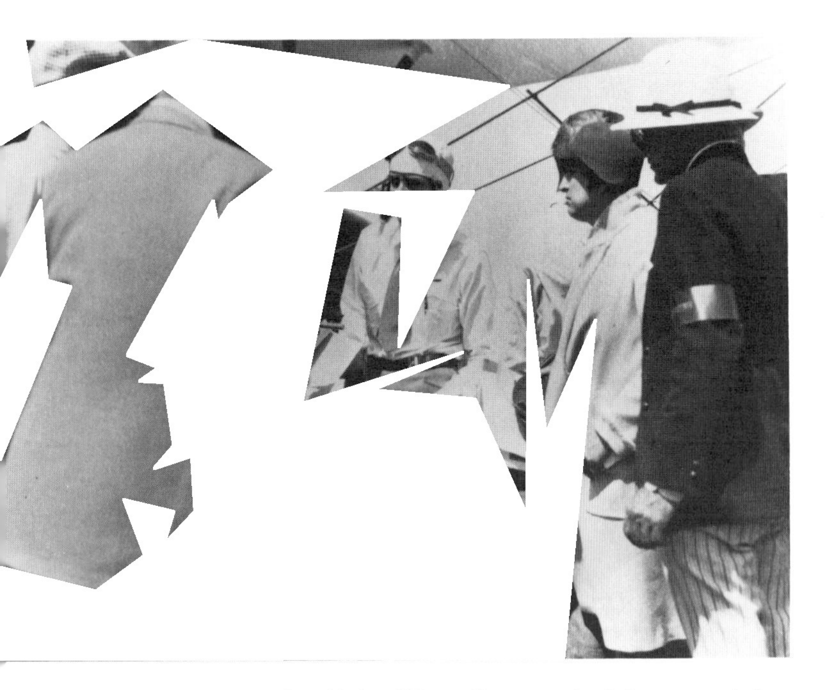

FIGURE 67.—Gladys O'Donnell in some final discussions before the start of the race. Elizabeth McQueen stands to her left. (Courtesy of H. Glenn Buffington)

cense. Her husband would follow the race, serving as her mechanic.

Claire Mae Fahy was also part of a flying duo. Her husband, Lt. Herbert J. Fahy, was well known in aviation circles, especially since he had established a new solo record on 29 May 1929. He followed his wife along the route of the race.

The sound of a pistol being fired in Cleveland, Ohio, was carried over the radio to signal the start of the National Women's Air Derby. One by one, 19 of the 20 entrants taxied down the runway in Santa Monica and took off. One pilot, Mary Haizlip, remained behind. She was still awaiting the arrival of a plane. Even before the start of the race Mary had run into problems. She left Tulsa, Oklahoma, en route to Santa Monica in a stock model American Eagle powered by the newly approved Wright J-6-7 radial engine. But while attempting to land in bad weather at Tucson, she dragged a wing and bent the propeller. There were no facilities for a quick repair at Tucson, so another American Eagle had to

FIGURE 68.—On 18 August 1929 at Santa Monica the Women's National Air Derby commenced. Thea Rasche, #61, is rolling. Lined up behind her in the first row are (left to right): Bobbi Trout, #100; Marvel Crosson, #1; Pancho Barnes, #2; Blanche Noyes, #3; and Louise Thaden, #4. In the back row are: Opal Kunz, #18; Ruth Nichols, #16; Margaret Perry, #11; Amelia Earhart, #6; and Mary von Mach, #5. (Courtesy of H. Glenn Buffington)

be sent for. A three-place, clipped-wing biplane equipped with a well-used Wright J-5 engine was delivered the following day, and Mary was able to get underway.

The first leg of the race to San Bernardino was only 65 miles, but they stopped there for the night. The air was thick with dust and negotiating a landing was not easy. Opal Kunz ground-looped, damaging one undercarriage strut, but it was repaired during the night. Amelia Earhart and Mary von Mach arrived late. Both had experienced engine trouble and had returned to Santa Monica for repairs before continuing their flights.

The first of innumerable chicken dinners was served, and was followed by a meeting where the pilots were informed of a route change. The stop at Calexico had been dropped, and they were to fly straight to Yuma, Arizona.

Before taking off from Santa Monica, Thea Rasche had shown Louise Thaden a telegram warning her to beware of sabotage. Although the threat

FIGURE 69.—Shared experiences of the grueling race built fast friendships among the pilots. Bobbi Trout poses here with Vera Walker. (Courtesy of Bobbi Trout)

FIGURE 70.—Before the takeoff from Douglas, Arizona the planes are checked. Amelia Earhart's Vega is in the foreground; the official's Vega is next to it. The airplane engines are shrouded in tarps to keep the dust out. (SI photo 79-10578)

was not taken too seriously, it was reported to race officials. But early the next morning, as if to prove that the telegram was no joke, came word that oil had inadvertently been poured into the gasoline tanks of two airplanes.

Departure from San Bernardino was set at 6:00 A.M., and by 8:00 A.M. the first pilots were approaching Yuma. The heat radiated from the ground and sand had drifted over the runway, making it difficult to see the landing strip from the air. Amelia ran off the edge, up-ended her plane, and buried its nose in the sand. The propeller was bent and she had to have another flown in from Los Angeles.

The race was not scheduled to resume until noon. As the pilots waited, word arrived that Claire Fahy had withdrawn from the race after a forced landing near Calexico. She believed that someone had sabatoged her plane by pouring acid over the wires. The investigation of the incident was inconclusive. Bobbi Trout and Thea Rasche were forced down before reaching Yuma, but both flew on after making repairs.

When Mary Haizlip, who had started a day late, arrived in the vicinity of Calexico after dark, she landed where she saw lights. Too late she realized that it was the wrong airfield. She had overshot her mark and ended up south of the border in Mexicali on the Baja Peninsula. Several hours passed before the red tape was cut and she was allowed to continue the race.

Phoenix, Arizona, was the next overnight stop, an oasis in the desert. Pancho Barnes arrived late. Having lost her way, she followed railroad tracks until she spotted a group of shacks. After landing, she discovered, much to her horror, that she was in Mexico. She lost no time in getting her plane into the air and heading back across the border. Only Marvel Crosson had not arrived by dinner time. There were rumors that she had crashed, but no definite news.

Tuesday morning the pilots headed for Douglas, Arizona. Opal Kunz and Chubbie Keith-Miller were forced to land just short of Douglas, both out of fuel. In Douglas word was waiting that Marvel's body had been found a short distance from her plane. She had apparently jumped but was already too low for her parachute to open. Details of her fatal accident were sketchy.

The remaining women continued their flight to El Paso, spurred on rather than daunted by Marvel's death. The competitors' feelings about the tragedy were later explained by Louise Thaden (1938:76):

> If your time has come to go, it is a glorious way in which to pass over. Smell of burning oil, the feel of strength and power beneath your hands. So quick has been the transition from life to death there must still linger in your mind's eye the everlasting beauty and joy of flight. . . . Women pilots were blazing a new trail. Each pioneering effort must bow to death. There has never been nor will there ever be progress without sacrifice of human life.

Not everyone was quite so philosophical. There were immediate calls for the race to be terminated. At El Paso the flyers and race officials were greeted with newspaper headlines proclaiming, "RACE SHOULD BE STOPPED," followed by the far from enlightened statement from a Texas oilman named Halliburton that "women have been dependent on man for guidance for so long that when they are put on their resources they are handicapped."[9] The response from the Race Committee as voiced by Frank Copeland, Derby Manager, was swift and to the point. "We wish officially to thumb our collective nose at Halliburton. There would be no stopping this race!"[10]

The deteriorating weather kept the flyers in El Paso overnight. Takeoff was scheduled for 6:00 A.M. the next morning, and Pecos, Texas, was the next stop. The airfield there was just a narrow strip that had been plowed out of the mesquite. To avoid running into the mesquite, Edith Foltz groundlooped her plane. The damage to the undercarriage of her Alexander Eaglerock was slight, and she was able to rejoin the race later when repairs had been made. Blanche Noyes made a delicate landing on one wheel at Pecos, and when she emerged from her crippled plane her face was blackened and her hands were scorched. The smell of smoke had forced her to land in the mesquite and she had discovered a fire in the baggage compartment. She speedily put out the blaze with the first thing she could lay her hands on, sand. In manuvering her plane out of the mesquite and into the air, she lost a wheel. It would be replaced and the wing repaired so that she could continue. Pancho Barnes had been forced to return to El Paso because of a leaky gas line, and her problems had not ended there. On landing at Pecos, she smashed her plane into an automobile that was parked too close to the runway. She emerged unhurt, but the damage to her plane was extensive, and she was forced to withdraw from the race.

---

[9] Newspaper clipping from the scrapbook of Bobbi Trout.
[10] Ibid.

FIGURE 71.—Some of the pilots gathered for photos while waiting for the fog to lift at Parks Airport, East St. Louis (left to right) were: Mary von Mach, Jessie Keith-Miller, Gladys O'Donnell, Thea Rasche, Phoebe Omlie, Louise Thaden, Amelia Earhart, Blanche Noyes, Ruth Elder, and Vera Walker. (Courtesy of Louise Thaden)

The flyers stopped at Midland, Texas, and then pressed on to Abilene. They wanted to make up for the time they had lost in the overnight stay at El Paso. After a short rest at Abilene they were off for Fort Worth, Texas. This was probably the most grueling day of flying they experienced during the race. Louise still held first place in the DW Class; Phoebe had a firm hold on first place in the CW Class. At Fort Worth Margaret Perry withdrew from the race. For two days she had been flying with a high fever, but now she was whisked away to a hospital. (For the next ten days she waged a battle against typhoid fever.) Ruth Elder had a close call en route. When she made an emergency landing in a farmer's field, she did not think about what animals were grazing there, at least not until they began approaching her bright red plane. Then her only thought was, "Please let them be cows."

There was a three hour stop at Tulsa, Oklahoma, and then it was on to Wichita, Kansas. Louise Thaden had family and friends waiting there; this was her "home ground." The pilots had left the intense heat and blowing sand behind, but one kind of bad weather had been replaced by another—stinging, driving rain. At Wichita, Louise maintained her lead, still closely followed by Gladys O'Donnell. Phoebe Omlie was still far ahead in her class.

The departure for Kansas City the next morning was delayed because of the weather, so to make up for lost time there was only a quick stop for lunch before the flyers pressed on for East St. Louis, Missouri. Neva Paris and Blanche Noyes had intentionally ground-looped during their landings at Park Airport, St. Louis to avoid overshooting the end of the runway. The damage was minor, and the planes were repaired by the time the pilots began arriving at the field the next morning. Mary Haizlip's plane was ready as well; the gas lines had been cleaned and cleared. The bent propeller blade on Gladys O'Donnell's plane had been straightened, so she, too, would continue. A final inspection of Louise Thaden's plane revealed that the magneto point had been filed, but it was replaced. Still, all the way to Terre Haute gas spewed from around the main gas tank cap. A check upon landing revealed that the

FIGURE 72.—Above, Louise Thaden at the finish. (Courtesy of Louise Thaden)

FIGURE 73.—Left, this tremendous shawl of roses was draped around Louise Thaden, and she immediately discovered that the flowers had not been "dethorned." The wreath was quickly removed, and hung over the propeller of her Travel Air. (Courtesy of Louise Thaden)

felt washer was missing.

In the mob scene at Terre Haute, the men on the gas trucks tried to fit the gas nozzles where they obviously wouldn't go, and during all the commotion someone unscrewed the oil-drain plug on Louise's plane. Fortunately, she noticed what had happened before getting underway, and the oil was replaced.

A stop at Cincinnati, Ohio, was added to the itinerary. With the change in terrain, navigation had become more difficult. Flat desert had given way to open plains, which in turn had become tree-covered hills and valleys. Mary von Mach, Opal Kunz, and Vera Dawn Walker doggedly kept up. They were the least experienced, but what they lacked in skill was made up by sheer determination and perseverance. Bobbi Trout was regrettably far behind, and was forced to land yet again with mechanical problems. When the rest of the flyers departed from Cincinnati, Edith Foltz had not arrived. There was concern because no one had heard from her; but at Columbus she caught up. She had lost her way between Terre Haute and Cincinnati. The race was nearing its end. The last lap would bring them to Cleveland, their destination.

On Monday, 26 August 1929, the last day of the race, takeoff was not scheduled until 1:00 P.M. This gave everyone a chance to rest, relax, and work on their planes. Following some work that had been done on her plane, Ruth Nichols decided to make a short test flight; but as she came in to land she failed to see a steamroller parked at the end of the runway and her Rearwin slammed into it. Ruth was unhurt, but her airplane could not be repaired in time for her to finish the race.

Of the 20 original entrants, 14 lined up on the runway at Columbus to fly the final 126 miles to Cleveland. The elapsed times of the leading fliers in the DW Class were close: Thaden had clocked 19:35:04 hours; O'Donnell, 20:32:25; Nichols (who was now out of the race), 21:15:45; Earhart, 21:17:37; Noyes, 23:45:51; and Elder 25:27:17. In the CW Class Phoebe Omlie held an undisputable lead, with an elapsed time of 22:14:32. One by one pilots were flagged off in order of their standings. A lot could happen in 126 miles, and it was anyone's guess who would be first across the finish line.

When her blue and gold Travel Air burst across the finish line, Louise Thaden did not immediately realize that she had won the Derby. The crowd surged around her and she found herself confronted with reporters and a battery of microphones. Gasp-

FIGURE 74.—Ruth Elder's friend and flying companion George Haldeman was waiting to greet her at the finish line. (SI photo 79-7510)

FIGURE 75.—Cliff Henderson, managing director of the National Air Races, talks with one of the starters. (SI photo 79-12248)

FIGURE 76.—Above, Ruth Elder and Cliff Henderson (center) at the National Air Races in Cleveland. (SI photo 79-5997)

FIGURE 77.—Left, Jessie Maude Keith-Miller took second place in the Women's light-plane closed course race. (SI photo 79-7509)

FIGURE 78.—Below, Elinor Smith, posing here with Bert White, also participated in the closed course events at Cleveland. (SI photo 81-12624)

ing for something to say, she finally managed (author unknown, 1974:175): "I'm glad to be here. All the girls flew a splendid race, much better than I. Each one deserves first place, because each one is a winner. Mine is a faster ship. Thank you."

And each one was a winner. One by one, 13 flyers followed Louise into Cleveland that day. Final positions for the DW Class were Thaden, O'Donnell, Earhart, Noyes, Elder, Paris, Haizlip, Kunz, von Mach, and Walker. The CW Class finished as follows: Omlie, Foltz, Miller, and Rasche. Trout flew in the next day.

What these women had accomplished by competing in and completing the Derby was by no means insignificant. They had provided the industry with yet another forum through which to test state-of-the-art aircraft. Most of the problems they confronted had been universal rather than particular to women. There were more mechanical failures of the aircraft than personal errors of judgment. Despite these problems, 15 of the 20 women completed the race course.

In the remaining days of the National Air Races there would be four closed-course events open to women only. A 50-mile race for DW and CW Class planes, and an Australian handicap race for both classes took place; the dead-stick landing event for women was canceled (see Appendix A). The public witnessed many thrilling moments as these women put themselves and their planes through their paces, but for those who had flown in the Derby this was all anticlimatic. For them there had never been nor would there ever be another race like that first National Women's Air Derby.

Just the Beginning

The 1929 National Air Races at Cleveland had drawn woman pilots from all across the country together. It was the first time that many of them had met face-to-face and had an opportunity to "talk shop" with other women. Licensed woman pilots were few and far between in 1929, and no matter how well they were accepted by male pilots and the communities in which they lived, there must have been moments of intense loneliness for the woman who was, quite literally, the only female pilot for hundreds of miles.

The consensus was that they needed to keep in touch with each other. Several women expressed a wish for an organization of women pilots through which they could share their experiences and encourage other women to take up flying, as well as work toward the general good of aviation. In the weeks following the races this feeling spread, but still there was only talk. Someone needed to provide the impetus to get such an organization underway.

It was Clara Trenchmann who finally set the plans in motion. She was not a pilot herself but was deeply interested in aviation and women's role in the field. Clara worked in the Women's Department of the Curtiss Flying Service at Valley Stream, Long Island, and since the previous September had been issuing a newsletter *Women and Aviation*. In it she reported on as many events as she could, her topics ranging from which women were taking flying lessons, who had qualified for licenses, and the latest trends in aviation, to tips and suggestions for safe flying.

On 1 October 1929, Clara Trenchman sent a memorandum to two Curtiss executives, Mr. Skinner and Mr. Mellem. In it she defined the need for a women pilot's association and suggested that four Curtiss demonstration pilots, Neva Paris, Frances Harrell, Fay Gillis, and Margery Brown, draft a letter to all licensed women pilots inviting them to meet at Valley Stream for the purpose of establishing such an organization.

The letter was sent out on 9 October 1929, setting the meeting for 2 November at 3:00 P.M. Twenty-six women arrived that day, though only four flew in. Despite the din in the hangar where they met, the foundations of the organization were laid. Neva Paris presided, Fay Gillis acted as secretary, and Wilma L. Walsh served as treasurer, collecting a $1.00 membership fee. They decided that the organization should be social as well as professional, and that membership would be limited to licensed women flyers.

But what would they call themselves? Almost every imaginable name was proposed and then rejected. Amelia Earhart finally suggested that the

FIGURE 79.—Some of the 26 women who gathered at Curtiss Field for the first meeting of the Ninety-nines: back (left to right), Neva Paris, Mary Alexander, Betty Huyler, Opal Logan Kunz, Jean David Hoyt, Jessie Keith-Miller, Amelia Earhart, Marjorie May Lesser, Sylvia A. Nelson, Dorothea Leh, Margaret F. O'Mara, Margery Brown, Mary Goodrich, Irene Chassey, "Keet" Mathews, E. Ruth Webb, and Fay Gillis; and front row, Viola Gentry, Cecil "Teddy" Kenyon, Wilma L. Walsh, Frances Harrel, and Meta Rotholz.

name be taken from the total number of charter members. Besides the 26 who attended the meeting at Valley Stream, 31 had sent their regrets. When the letter had been mailed on 9 October, however, a complete list of names and addresses was not available. Amelia Earhart and Neva Paris were appointed to send out a second letter, this time to include all those woman pilots who had been missed by the first letter. One hundred twenty-six women were invited to attend the next meeting at Opal Kunz's home in New York City. On 14 December 1929, 24 attended, but many more had responded. When the final tally was made they had their name—The Ninety-nines (see Appendix B).

It had been a dramatic decade for women in aviation—one dedicated to establishing their credibility. In 1919 the American public had viewed women interested in aviation with skepticism. With few exceptions airplanes and flying belonged to a man's world, for there was a certain mystique about flying which implied that superhuman qualities were required. Yet women were as fascinated by the machines and as exhilarated as any man when they took to the air. They were confident that they could perform as well as any man, and they set out to prove this.

Conclusion

In the 1920s women were a significant force for the progress of aviation. They were among the most daring of the barnstormers, risking their lives routinely in their quest for new and more exciting stunts. Others strove to promote flying as a safe and convenient way to travel. They tested and demonstrated new planes. They carried passengers and gave flying lessons. Women used a variety of means to help aviation develop as a technology, an industry, a business, and a sport. They set new records, constantly testing the limits of the planes and of their own abilities. They raced, both for the joy of the sport and to prove their competence as pilots. They established a category for women's records and founded their own pilot's organization, which is still active today.

By their active participation women helped aviation come of age in this country. Their courage and determination as they struggled to meet the challenges of being pilots and to overcome the barriers they faced as women caught the imagination of the country. Their contributions helped to bring aviation into its present place as vital industry in modern society.

Appendix A

National Air Races
Cleveland, Ohio, 24 Aug–2 Sept
Scheduled Closed-Course Events Open to Women

Event No. 1
Ladies CW Class Race

(1) Open to all type planes powered with motor or motors having a total of less than 510 cubic inch displacement.
(2) 10 laps of 5-mile course.
(3) Prizes, $1000.00: 1st $500.00; 2nd $300.00; 3rd $200.00.

	Pilot	Plane	Motor	Speed (mph)
1	Phoebe Omlie*	Monocoupe	Warner	112.37
2	Keith-Miller	Fleet	Kinner K-5	98.73
3	Lady Mary Heath	Great Lakes	American Cirrus	96.17
4	Blanche Noyes	Great Lakes	American Cirrus	85.12

* Phoebe Omlie was disqualified for cutting a pylon, and the prize went to Jessie Maude Keith-Miller

Event No. 28
Ladies DW Class Race

(1) Open to all type planes powered with motor or motors having piston displacement of more than 510 and not more than 800 cubic inches.
(2) 12 laps of 5-mile course.
(3) Prizes, $1250.00: 1st, $625.00; 2nd, $375.00; 3rd, $250.00.

	Pilot	Plane	Motor	Speed (mph)
1	Gladys O'Donnell	Waco Taper Wing	Wright J-6	137.60
2	Louise Thaden	Travel Air	Wright J-5	131.43
3	Blanche Noyes	Travel Air	Wright J-5	127.77

Event No. 29
Dead Stick Landing Contest for Women Pilots

(1) This contest will be held on three different days of the races, 29 August, 31 August, and 2 September.
(2) Prizes, $175.00 each day: 1st, $100.00; 2nd, $50.00; 3rd, $25.00.

CANCELLED

Event No. 30
Australian Pursuit Race for Women
(To be held Friday, 30 August)

(1) Open to all types of planes with motor having piston displacements of not less than 275 and not more than 800 cubic inches.

(2) Using the known speed of the various airplanes entering and the length of the course as factors, the time that each airplane should require to fly the prescribed course will be worked out, the faster planes will be handicapped, and the planes so started that they should finish at exactly the same time.

(3) When one plane passes another, the plane passed must drop out of the race. If more than 15 entrants are ready to start, semi-finals will be run.

(4) 12 laps of 5-mile course.

(5) Prizes, $1250.00: 1st, $625.00; 2nd, $375.00; 3rd, $250.00.

	Pilot	Plane	Motor	Speed (mph)
1	Gladys O'Donnell	Waco Taper Wing	Wright J-6	138.21
2	Thea Rasche	Gypsy Moth	Gypsy	97.31
3	Frances Harrell	Gypsy Moth	Gypsy	112.24

Event No. 31
Australian Pursuit Race for Women
(To be Held Sunday, 1 September)

(1) Open to all type planes with motor or motors having piston displacements of not less than 275 and not more than 800 cubic inches.

(2) Prizes, $1250.00: 1st, $625.00; 2nd, $375.00; 3rd, $250.00.

(3) Distance: 12 laps of a 5-mile course.

	Pilot	Plane	Motor	Speed (mph)
1	Thea Rasche	Gypsy Moth	Gypsy	99.72
2	Louise Thaden	Travel Air	Wright J-5	136.83
3	Gladys O'Donnell	Waco Taper Wing	Wright J-6	137.63

Appendix B

Charter Members of the Ninety-nines

Mary C. Alexander
Mary Bacon
Barbara W. Bancroft
Bernice C. Blake
Ruth T. Bridewell
Margery H. Brown
Myrtle Brown
Vera Brown
Thelma Burleigh
Myrtle R. Caldwell
Ruth Elder Camp
Mildred Helene Chase
Irene J. Chassey
Bonnie M. Chittenden
Marion Clarke
Margaret Perry Cooper
Helen V. Cox
Jean Davidson
Jane Dodge
Marjorie Doig
Amelia Earhart
Thelma Elliott
Frances Ferguson
Sarah Fenno
Adeline F. Fiset
Phyllis Fleet
Edith Foltz
Ila Fox
Viola Gentry
Betty Huyler Gillies
Fay Gillis
Phyllis M. Goddard
Mary Goodrich
Melba M. Gorby
Geraldine Grey
Candis I. Hall
Sacha Peggy Hall
Ruth E. Halliburton
Frances Harrell
Lady Mary Heath
Jean D. Hoyt
Katherine F. Johnson
Angela L. Joseph
Mildred E. A. Kauffman
Betsy Kelly
Madeline B. Kelly
Teddy Kenyon
Cecelia Kenny
Florence E. Klingensmith
Opal Logan Kunz
Eleanore B. Lay
Eva May Lange
Jean LaRene
Dorothea Leh
Marjorie Lesser
Ethel Lovelace
Lola L. Lutz
Edwyna McConnell
Retha McCulloh
Helen Manning
Olivia Matthews
Jessie Keith-Miller
Agnes A. Mills
Sylvia Anthony Nelson
Ruth Nichols
Mary N. Nicholson
Blanche W. Noyes
Gladys O'Donnell
Margaret F. O'Mara
Phoebe Fairgrave Omlie
Neva Paris
Peggie J. Paxson
Achsa Barnwell Peacock
Elizabeth F. Place
Lillian Porter
Thea Rasche
Mathilda J. Ray
Meta Rothholz
Gertrude Catherine Ruland
Joan Fay Shankle
Hazel Mark Spanagle
Ruth W. Stewart
Marjorie G. Stinson
Mildred Stinaff
Dorothy L. Stocker
Louise M. Thaden
Margaret Thomas
Nancy Hopkins Tier
Evelyn "Bobbi" Trout
Esther M. Vance
Mary E. Von Mach
Wilma L. Walsh
Vera Dawn Walker
Ruth E. Webb
Nora Alma White
Nellie Z. Willhite
Margaret Willis
Josephine Chatten Wood
Alberta B. Worley

References

Adams, Jean, and Margaret Kimball
1942. *Heroines of the Sky.* Garden City, New York: Doubleday, Doran & Co.

Author unknown
1919a. *The Ace* (September), 1(2):30.
1919b. *The Ace* (October), 1(3):16.
1919c. *New York Times* (22 October), page 17.
1921. Editorial. *New York Times* (7 June), page 16.
1922. *New York Times* (9 October), page 1.
1924. *Aeronautical Digest* (July), 4(1):7.
1927a. *New York Times* (10 October), section 2 page 3.
1927b. *New York Times* (13 October), page 1.
1927c. *New York Times* (14 October), section 3, page 2.
1927d. *New York Times* (17 October), page 2, column 1.
1927e. *Popular Mechanics* (December), pages 885–889.
1928a. *New York Times* (4 June), section 2, page 3.
1928b. *New York Times* (8 June), page 24, columns 3, 4.
1928c. *New York Times* (13 June), section 1, page 2.
1974. The National Women's Air Derby. *Aviation Quarterly,* 1(3):175.

Ball, Daisy Elizabeth
1923. Women's Part in Aviation. *Aeronautical Digest* (July), 3(1):25.

Barker, Ralph
1971. *Verdict on a Lost Flyer, the Story of Bill Lancaster.* New York: St. Martin's Press.

Briddon, Arnold E., and Champie, Ellmore A.
1966. *Federal Aviation Agency Historical Fact Book: A Chronology 1926–1963.* Washington, D.C.: Federal Aviation Agency.

Brown, Margery
1929. What Men Flyers think of Women Pilots. *Popular Aviation and Aeronautics* (March), 4(3):62–64.

Buffington, H. Glenn
1974. Flair for Flight. *Ninety-nine News* (March/April), pages 10–12.

Caidin, Martin
1965. *Barnstorming.* New York: Duell, Sloan, and Pearce.

Corn, Joseph J.
1979. Making Flying Thinkable: Women Pilots and the Selling of Aviation, 1927–1940. *American Quarterly* (Fall), 31:556–571.

Dwiggins, Donald
1968. *The Barnstormers: Flying Daredevils of the Roaring Twenties.* New York: Grosset & Dunap.

Earhart, Amelia
1928. *20 Hrs. 40 Mins.* New York: Grosset & Dunlap.

Gentry, Viola
1975. *Hangar Flying.* Printed in Chelmsford, Massachusetts.

Heath, Mary
1929. Women Who Fly. *Popular Aviation and Aeronautics* (May), 4(5):32–34.

May, Charles P.
1962. *Women in Aeronautics.* New York: Thomas Nelson & Sons.

Marcone, Theodora
1924. *Aeronautical Digest* (July), 4(1):86.

Nichols, Ruth
1957. *Wings for Life.* Philadelphia: Lippincott.

Oakes, Claudia M.
1978. United States Women in Aviation through World War I. *Smithsonian Studies in Air and Space,* no. 2.

Orr, Flora J.
1935. Southern Personalities: Phoebe Fairgrave Omlie. *Holland's, the Magazine of the South* (September), page 14.

Stilwell, Hart, and Slats Rodgers
1954. *Old Soggy No. 1, the Uninhibited Story of Slats Rodgers.* New York: Julian Messner. [Reprint, 1972, New York: Arno Press, Inc.]

Thaden, Louise
1938. *High, Wide and Frightened.* New York: Stackpole & Sons.

Wells, Fay Gillis
[1945]. *The 99 Club, an International Organization of Licensed Women Pilots.* 32 pages [Washington, D.C.].